The Science Of God Volume 4

R Lindemann

Aleph Publications
Wisconsin, USA

The Science Of God Volume 4
Day Six - Evolution versus Man - In Our Image
Copyright 2021 - R Lindemann ©
All Rights Reserved. Published 2023

Aleph Publications
Manitowoc WI 54221

All rights reserved. No part of this publication may be stored in a retrieval system, reproduced, or transmitted in any form, electronic, mechanical, photocopying, recording or other, without first obtaining the written permission of the copyright owners and the publisher.

Paperback Edition
ISBN13: 978-1-956814-30-9

33 32 31 30 29 28 27 26 25 24   2 3 4 5 6

Disclaimer

All information, views, thoughts, and opinions expressed herein are those of the author(s) and are being presented only for your consideration and should not be interpreted as advice to take any action. Any action you take with regard to implementing or not implementing the information, views, thoughts, and opinions contained within this published work is your own responsibility. Under no circumstances are distributor(s) and/or publisher(s) and/or author(s) of this work liable for any of your actions.

Anyone, especially those who have been victim of misdirected explanation and understanding, may be best served seeking wise counsel before deciding to implement any information, views, thoughts, opinions, or anything else that is offered for your consideration in this work. All information, views, thoughts, and opinions in this work are not advice, directive, recommendation, counsel, or any other indication for anyone to take any action. All information, views, thoughts, and opinions offered herein are offered only as suggestions for your personal consideration, which is done of your own free will. Your life is your own responsibility; use it wisely.

Any use of trade names or mention of commercial sources is for informational purposes only and does not imply endorsement or affiliation.

Please note that most of the items in quotes in this book are from various versions of the Bible and some quotes may have been paraphrased.

Dedication

To Father Adam and Mother Eve who forged a path of courage for all of us in the face of their utter innocence. It is impossible for all of us, your children who follow your path, to understand what it must have been like to be in paradise with God living in absolute innocence. You must have experienced true peace and calm, but then total fear as you began to become aware of *good* and *evil*.

Many of us today think, "How could they have followed Satan's lure and eaten the fruit of the tree of Knowledge of Good and Evil after having been warned by God?" But we never experienced total innocence, total peace, and total trust. Nor have we ever experienced not knowing between good and evil as aware adults or even as children.

It must have been horrifying for both of you when your eyes were opened and you realized that Satan had lured you into doing evil by stealing away your innocence. We now understand that God knew that this would likely occur because of the very fact that you had perfect innocence, so pure and beautiful and childlike. You had innocence far more pure than even that of a two-year-old.

So Father Adam and Mother Eve, we thank you for your courage. And this book is dedicated to you in recognition of your parental authority over us—your offspring. Blessings to you and to the Creator God who Created you! Thank you!

Contents

Chapter 1
Conceived in Our Image 1
    The Concept of Mankind 3
    Discrediting the Bible and Science 4
    From Our Perspective 5

Chapter 2
"I Think Therefore I Am" 7
    Proving Existence 8
    Godless Tactics 12
    Incorrect Assumptions 15
    An Ability to Reason 16

Chapter 3
Man - The Higher Creation 19
    Higher Mind 21
    Human Ingenuity 24
    All Kinds of Things 26
    Mankind 28
    Made from Slime 29
    Did Adam Have a Belly Button? 30

Chapter 4
A Way to Perfection 37
    Evolution Absurdity 38
    Can We Be Too Precise? 39
    More Than Two of Each 41

Chapter 5
The Breath of Life 43
    Not Believing Anything 43
    At First Breath 44
    Adam's First Breath 46

Chapter 6
Woman - The Perfect Companion 49
    A Pristine World 49
    From a Rib 51
    Construction Differences 54
    Choice of Breeding 55

Chapter 7
Multiply by Dividing 59
    Repetitious 60
    Definition of Terms 62
    Defining a "kind" 63
    Hebrew, Latin, and Greek 64

Chapter 8
Forming the Vessel 65
    By Design 66
    Versions of Slime 67
    Replication Errors 68
    Eukaryotes, Prokaryotes and More 70

Chapter 9
Preserving Creation for the Long-Term 71
The Bones of the Matter 72
Phylogeny 73
Long Age Layers 75
Written Preservation 76

Chapter 10
What We Are Committed To 79
Don't Be Fooled 79
Debatable Points 81
Animals Versus Humans 83

Chapter 11
A Seed of an Idea 85
Off the Rails 86
Lack of Definition by Too Much Definition 87
Thus Sayeth the Lord 89

Chapter 12
Guided by the Layers 91
Rants Are Questions 92
Our Found Species 93
Race, Creed, and Color 95

Chapter 13
Science is Fluid 97
Success of Moving Theories 98
The Evolution Task List 99
Pop-Science 100
Real Science 100

Chapter 14
What is "Life" 103
Gene Sequencing 104
Genetic Additions and Deletions 105
Making Life In a Laboratory 106

Chapter 15
Peer Review On the Same Page 109
Observations versus Conclusion 110
Know It All 112
Progression of Evolution 113
What is "Evolution"? 115

Chapter 16
Ten-Thousand Scientists Can't Be Wrong 117
Being Specific About Defining Terms 118
Getting Published 119
Creating Theories 120
Our Ingenuity 122

Chapter 17
Our Attempts to Share 123
Becoming Civil 124
Bible Abroad 126
Same as 3500 Years Ago 126

Chapter 18
Believe What You Will **129**
- Filling Our Desire 129
- Arrogant Statements 130
- A Fishy Story - Hook, Line, and Sinker 132
- Micro Evolution 133
- Is a Fetus Really "Just a Clump of Cells"? 133

Chapter 19
Preserved for Safe Keeping **139**
- Separation of Ages 140
- Foot Prints Written In Stone 141
- Macro-Evolution 143

Chapter 20
What Does the Data Indicate? **145**
- Our DNA Data 146
- A Scientific Method 147
- The Scientific Method 147
- Is Biblical Data Innocent? 148
- Reading the Data 149

Chapter 21
You Becoming You **153**
- Skeletal Finds 154
- Facts Are Facts 155
- Some Facts Are Not Facts 156

Chapter 22
Science and The Bible **159**
- Who Gets to Define "kinds"? 161
- Believers versus Christians 162
- Is the Bible Scientific? 164
- The Spirit of God 164

Chapter 23
Ask and You Shall Receive **167**
- Expectations 168
- Earliest Primates 171
- Believe what You Will 173

Chapter 24
Final Analysis **175**
- Intelligence of the Ancients 176
- Dangers of "science" 177
- Wonders of Science 178
- Scientific Salvation 179
- We Are Significantly Unique and Special 183

Acknowledgements

A great many thanks to all who study origins and have contributed to the collective knowledge of mankind. Our collective data is there for all of man to study and understand. Thank you for your research, your efforts, and your thoughts!

I would also like to take this opportunity to thank all who have been a part of the distribution of the several *The Science of God* volumes, from the lumberjack who cut the trees to the driver who hauls them. From the oil field worker who drills for the oil that goes into the truck to haul the trees that make the paper for the shipping bill that the driver carries to the mill for the office worker who handles the paperwork with the ink from the person who works at the ink company who drives their car to the garage where the technician checks the engine made by auto workers who buy tires from the company who employees the wife of the wood hauling trucker driver, the wife who works at the software company who helped write the software that was used to Create the manuscript for books like this.

As you can see, the list of people mentioned above doesn't even begin scratch the surface of all of the wonderful people who are involved in our daily lives that we never even realize have taken part in our day. There are too many people to thank even just by broadly mentioning the basic industries as was briefly done in the previous paragraph. In truth, it would truly fill volumes of books to even begin to acknowledge everyone who made your reading this possible. In fact, it is possible that even you yourself contributed in some remote way to the ability for a book such as this to reach you. This is also true of those noble women who stayed home to bring up their children, children who now work in the many fields utilized in the Creation and distribution of a book such as this. So thanks to those of you who are reading this for doing your part in life, and thanks to all of those who show up for work every single workday to make this world a better place for all of us!

Introduction

In our "scientific" world, we are faced with a deluge of "compelling" scientific data. If you have gone to public school, you will have had more than your share of exposure to some of this data, but often in a somewhat indirect manner. The subtle indoctrination regarding evolution being true, and the Bible's Creation account being false, is really quite overwhelming when you look into that aspect of modern education. And when it comes to the arrival of mankind on Earth, we end up splitting hairs and misinterpreting a lot of information regardless of which perspective, religious or scientific, we are viewing the information from.

From an evolutionary science perspective, nothing really matters, and ultimately there are no rules or guidelines. However, from a Biblical Creation perspective, "man" is the pinnacle of Creation, and if man or animals could interbreed with other kinds it would pervert their kind. The desire for our own kind that is in good physical form is a protection for humanity against selecting a path that will lead to the perversion of the ultimate design of man. So long as men and women always seek women and men that they are attracted to, the world will stay diverse and humans will remain pure and prolific. If this could ever be breached, then humans will have tainted their origins and will no longer be pure and in harmony with the Creator's original design.

There is much about this subject that the world is not prepared to hear and will reject. Some of the myths of old have some basis in truth, and we have been misled into believing that certain crossbred animals never existed. We fail to believe what ancient writings state and is sometimes witnessed in the fossil record. We continue to embrace the erred evolution view that disregards all things that do not abide by the rules of evolutionism whose mantra is, "Throw out all evidence that does not support the theory". Doing so is dishonest and is a lie. When we fail to properly analyze the presented evidence, we then each have deceived our own selves.

From a Biblical perspective, we were Created unique from other kinds of beings, namely animals. And to violate that is a "sin", which is to say that it is void of the desire of God the Creator. That point is very important for all of us to understand. If a Creator does in fact exist, we are ultimately judging ourselves by our choices, possibly resulting in separating ourselves from this alleged Creator. If you buy into the Creation argument or the evolution argument, you then have made the choice as to what you believe. Did you get that? **You** have made the choice. And here is an important point: It may very well be that neither of these two standard belief sets are correct. This means that it is possible that no matter which choice you make, you might be wrong and ultimately be liable for your error—especially if you misguide others using that error.

As you read on, try to remove all biases and read from a fresh perspective, and then analyze the information that you read and see how it compares to what you may have thought before reading this book. Will the information that you read in this book reconcile with science? Will it reconcile with the Bible? And, will it reconcile with reality? You be the judge as you read on and consider the points presented in this volume of *The Science of God*. It is important to note that in this book when the term "man" is used it is typically referring to what we call "mankind" rather than "man" or "woman".

Chapter 1

Conceived in Our Image

The question everybody wants to know or prove is: Are humans "Created", or did we "evolve"? That question has been plaguing mankind for hundreds of years, and more than likely, it has been asked for thousands of years through most of recorded human history. Most people are seeking actual answers to the question rather than the petty arguments and name calling we all too often hear. When we're passionate about our beliefs and are frustrated because we are unable to answer all of the questions we're asked, that's typically when voices are raised and mockery ensues, but all to no resolution to our actual questions. Are **you** willing to hear other people's thoughts on the matter? Or is your mind closed and it is a settled point of "inarguable fact" in your mind?

Since we are discussing Creation in this book, we must to first make the assumption that a Creator exists and analyze that and then dissect parts of Genesis One to see if it is scientifically viable that a "God" Created mankind, or if Genesis One from "The Bible is nothing more than mere fairytales". But let's not stop there; let's also take a look at the science side to see if it is real, or if it is "mere fairytales".

To start this quest to understand whether we evolved or were Created, we first have to find the starting point and ask: What are our positions' parameters of the "Creation" of mankind? Will we allow any transitional adaption from animals? Or was mankind made from the very start just as we witness ourselves today? This is a very important detail to nail down in the *evolution*-versus-*Creation* debate, because if our foundations of discussion are moving targets, then any arguments or theories built upon those foundations become worthless while in debate. We see this problem on both sides of the debate where they will take an initial theory and build their house of cards upon it. Then later, their initial foundation is forfeited on which their theory is built, yet they will still hold to the theory even though the foundation is now gone, which is what supported the theory to begin with. This sort of disassociated reasoning is of little use in finding out how it all actually occurred and how mankind truly came to be.

We should have no problem with people standing by a theory that they have concocted, but then they should at least be able to explain that theory and answer **all** of the questions that people have regarding the theory. In effort to clear the fog around the *evolution*-versus-*Creation* question, in this book we are going to take two specific fundamental positions:

First is the Creation perspective: The Creator God made man fully formed from the very start.

Second is the evolution perspective: We evolved without any intelligible guidance whatsoever, and only natural forces guided evolution to make us what we are today, and there is no "God".

So, either God made humans as we are today from the very start, or we are the result of a long, long path of evolutionary adaption. Which is it? Let us investigate this further.

The Concept of Mankind

The theory of evolution is quite well-known and does not involve any sort of discerning guidance to form the various species. This book focuses quite a bit on the "Created" perspective as it dissects the Genesis One day-six text. When we take the Biblical position of mankind having been Created, we have to ask ourselves: Did God specifically intend to Create mankind, or were we accidental and Created through the functions of evolution by God's command?

There are many followers of the Bible who have decided to accept the possibility that God only Created the needed materials, but then it was through evolution that mankind came to be and that the first male and female primates of discernable intellect were the Bible's Adam and Eve. This position is often utilized as an attempt to try to reconcile science with the Bible's Creation account when they cannot answer the tough questions. But this is contradictory and is one of those moving targets previously mentioned. I see this as a cop-out if you are going to claim that the Bible is "true". Additionally, there are those who believe that each of the six days of Creation were twenty-four-hour days like we experience today, and that all things were Created within a single seven-day week. But as discussed in the previous volumes of *The Science Of God*, that is not realistic or scientific, nor is it Biblical.

Since the Bible is the standard around which western culture is formed, and it is the "God" mentioned in the Bible that we believe in and pray to, then for us to make any claim that is not strongly implied in the Creation text is disassociated thinking and reasoning. If we believe the Bible to be "true" but then acquiesce to the scientific evolutionary model in the face of what the Bible actually says, then we are entertaining two disassociated lines of thought that cannot realistically work together as if they are one.

Since it is the Bible's Genesis One text being discussed, let's consider how the concept of "man" was conceived. When we humans make something, we first must have some idea of what we are going to make. After that, we then proceed to gather needed resources and we begin our Creative efforts. The same would be true of any being who has the intellectual capacity to Create anything.

Interestingly, mankind was Created on the same day that the beasts of the earth were Created. Here is the part of the Genesis One text where "man" comes into the picture: "And God said: Let the earth bring forth the living creature in its kind, cattle and creeping things, and beasts of the earth, according to their kinds. And it was so done. And God made the beasts of the earth according to their kinds, and cattle, and every thing that creepeth on the earth after its kind. And God saw that it was good. And he said: Let us make man to our image and likeness: and let him have dominion over the fishes of the sea, and the fowls of the air, and the beasts, and the whole earth, and every creeping creature that moveth upon the earth. And God Created man to his own image: to the image of God he Created him: male and female he Created them." Here we see in the text that God Created "man to his own image", yet we biologically share so much with the "beasts of the earth" that were Created earlier on the same day-six-event. So here we have two interesting points of consideration: According to Genesis, we share that same Creation day as the animals, yet we of mankind are made in the image of the Creator. In the Biblical sense, do those two points have any significance?

Discrediting the Bible and Science

Debates on *evolution* versus *Creation* are often done from negative perspectives in an effort to discredit the opposing perspective. There is nothing wrong with questioning the opposition's theories, but when both sides of the debate are able to legitimately discredit the opposition's theory, then what exactly is it that they have built their theories upon?

In the eyes of each particular critic, from either side, they attempt to discredit the other side because parts of that opposing theory do not reconcile within the theory itself. This problem is true of both sides, and both sides can be quite easily dismantled by most skilled articulators. While some of the points that the detractors make regarding discrepancies in opposing theories may indeed be rightfully pointed out, this does not disqualify the remainder of the information in either of the theories. Tearing someone else's theory down to make your own theory look more viable is not a good way to conduct science. It doesn't matter which side someone is on in this debate, because there are plenty of errors in everyone's thinking for everyone to have their full share of error. Any theorist can have a following of believers, and regardless of which side you debate from, you will likely have enough people who will rally to your side to belligerently shout your messages from their little soapboxes. But no matter what you demand to be true, does not make it true. Only what is true is true.

There is no shortage of blind scientists, or blind Bible believers. And what is worse is that often the multitude of errors in either sides' theories tends to disqualify, in our mind, any actual good and accurate information that either side presents in the mix of inaccuracies. Then to add to this problem, often the arguments stand on the stupidity of the opposition who take an unyieldingly ignorant approach in their beliefs. Just because the other side is wrong does not make our own side right. We see this sort of discrediting all the time when looking into this subject, and it does very little to get onlookers interested enough to form their own well-reasoned thoughts.

From Our Perspective

We have a tendency to view everything from our own current perspective, believing that things have always been as we see them today, but this is neither proper nor true. Often when something is written, it is written for a reason, which is especially true with regard to ancient warnings. Our world-culture is generally quite civil, but it has not always been this way regarding mankind. Culture has been proven to repeatedly ebb and flow from robust to destructive. We see this in ancient writings and in archeological digs. The past several hundred years have been a time of robust flourishing for mankind. We have been so productive and innovative to a point where we have actually been exploring the heavens, yet we are unable to come to any agreement as to how mankind came to be.

On the evolution side of this topic, it is understandable that people would see things only from a human perspective, because in general, they do not believe in God or a Creator who used reason to Create all that we see. However, on the Bible believers' side, it is a bit troubling with those who use only a human perspective when trying to understand how mankind came about.

We can take the blind-faith position and simply say "well, God did it" and then make some broad statements about how we imagine that happened, or we can hijack the evolution theories and falsely attribute those to "God's handiwork". But it appears that we seldom or never try to see the situation from **God's** perspective. This is true of both sides, but the Creation believers are more culpable in their negligence in the matter of *perspective*. Let's all try looking at this debate from a Creator God perspective for a change.

Chapter 2

"*I Think Therefore I Am*"

I would like to dive directly into the question of mankind having been Created, but since mankind is allegedly "Created" it quickly brings to the table the question of—Who or what exactly was it that caused mankind to be? However, first we need to understand that there is really no such thing as man-*kind*, but rather, it is simply to be stated as "man", but more on that later. Now, since Creating is an intellectual activity, we must dispose of the "*what*" and focus on the "*who*". When pondering these origins topics, whether it be questioning how we came to be, or how the Universe came to be, I like to take people back to the very beginning—back even further than Genesis chapter one verse one.

Ask yourself this: No matter how things came to be, how did it get there? Let's say that everything did come from a "big bang" which theoretically started as a point so infinitely small that it would be scientifically undetectable, we still have to ask how it got there. Size is a relative aspect of existence, so it does not matter how big or how infinitely small something is, *it* still is there. So if your position is big bang, then ask yourself: How did the pre-bang point of matter get there? Was *it* always there, and if so, then how did it initially get there? Saying that something was always there is a very unstable and irrational position and is really quite absurd as a scientific perspective—it is, technically, a very unscientific perspective that defies all scientific logic and reason.

But let's also take a look at this supposed "Creator". Do we imagine that God always existed? If so then, how did God get there? To assume that the Creator, or "God", always existed is no less irrational and illogical than what the big bang believers do. So, how did God get there? Most Bible believers make the foolish blind-faithed assumption that "God always existed". Believers must defend that position, but it defies logic so they are then forced to admit that they use blind faith, and their blind faith is no different than that of those who believe in the "big bang". So how did it get there? Or how did God get there? It doesn't matter if it is the pre-bang infinitely small point of matter, or if it is a Creator who has intellect, how did any bang substance or God get there?

Proving Existence

Asking how the first of anything came to be is somewhat of a tough topic to tackle because our minds logically gravitate towards some sort of intelligent intervention. With big bang followers, they tend make all matter so infinitely small that they can claim that it would be scientifically undetectable, therefore it does not exist, and from it everything came to be due to the big bang as it expanded. Again, it is either there or it is not, and to say that it was so small that it is not scientifically detectable is a cheap cop-out of an explanation that is built on absolute blind-faith.

Then there are those who believe that there is a Creator and they quickly run into problems with Creation when they are taken to task on how the Creator got there. With Creation, people simply take the position that "God Created everything", and in general, for many people that will suffice as the answer. And if God does in fact exist, then that is a completely logical answer, though it does invoke more questions from those who are looking deeper.

We have to use logic, and in order to discuss this entire topic as to whether it was Creation, or evolution, we truly need to know how God got there, or at least have a very plausible theory if we want to be able to have a rational discussion. If we are, in fact, Created, then we, as the Creation "man", (That is to say *mankind*) deserve to have an answer to the question of how the Creator came to be. Can we reason through this question to arrive at a logical and reasonable answer?

Consider this: Prior to the universe existing, regardless of how it came to be, *nothing* was there beforehand-**N o t h i n g**! This is a very tough thing for most people to mentally grapple with. We can't just assume that there was a bunch of space debris floating around that formed all that we see in the heavens, because then we have to answer where that debris came from. The logical answer here is—*nothing* existed beforehand. Similarly, if Creator God was not always there, then how did this Creator get there? Was there something or someone there beforehand that caused the Creator to be? Logically we must answer "no", otherwise we will catch ourselves in a never-ending cycle to infinity of a Creator's Creator's Creator's Creator, and on that would go for infinity.

At some point, we have to acknowledge that there was *nothing*–absolutely nothing! And here is where both positions get trapped in their infinite succession of irrational thinking. The big bang ideology has a problem that cannot be resolved in any way when trying to answer how the infinitely small point came to be, but depending upon your view of a Creator, the deliberate Creation view does not share that same inescapable problem. But it does if you are going to illogically claim that the Creator always existed. Claiming the Creator always existed is extremely illogical, especially since we are supposedly Created in the image of that Creator who used logic to Create.

This is where we have to examine ourselves. What are **you**? Are **you** your body? Or is it that your body is where **you** reside? The reason this is being brought into this discussion, is that if we are going to believe in a Creator, and if we believe that we are Created, then we must begin to understand what we are and how our thoughts are formed. Are thoughts tangible that they can be detected in anyway? Maybe. There have been plenty of scientific observations that when we think certain kinds of thoughts we can detect electromagnetic signals that are being emitted from the brain. But what makes those thoughts occur? This is an area of great contention and it is ultimately a question of whether or not a soul exists. This area of study is a bit complicated and speculative because some things are simply not detectable by any current scientific means or instruments.

There is no shortage of stories of people who died in hospitals and came back to life who claim that they could see what was occurring in the room where their body lay dead. Are these experiences their imaginations run wild, or did they really have the "out-of-body-experiences" that they claim to have had?

For those who experienced death in that way, there is nothing that you can do to convince them that they did not experience what they claim that they experienced. For them it is as real as you are right now as you read this book. So, this comes down to us being able to accept that there is such a thing as a soul. If we could somehow prove, in a scientific way, that we are in fact each a conscious soul, it would devastate the current scientific perspective on both big bang and evolution, because since a soul is undetectable and is able to think and reason—it is made of nothing, yet it exists. There currently is no known physical substance to a "soul".

Our "conscious" is truly an interesting and deep concept to consider. If our soul, or conscious, is not made of any substance, then we can assume the same is true of the Creator. Thus, the Creator is not some old man in the sky and is not made of anything tangible. The Creator is nothing more than ultimate conscious. Here's where this get interesting. To remain logical and not trap ourselves in a Creator's Creator's Creator's Creator... etc., we must admit that the Creator is Self-Created. That is to say that the Creator became aware and developed conscious, or maybe when conscious developed, then the Creator *became*. I realize that thought is a bit deep, but take some time to consider it.

Logically, the Creator had to have become conscious without any intervention from anything or anyone, which might sound fantastical, but it is far more plausible than any other theory previously mentioned from either side. But let's take a look at this a bit further. Since there was **nothing** before everything came to be, and since a soul or a conscious when viewed from a scientific perspective is undetectable and therefore is not made of anything and does not exist, we can then safely assume the same is true of the Creator. The Creator is conscious not made from anything. Can this be true? Is the Creator not made? Here is what we have to realize about this supposed "Creator": The Creator had *infinite* past to become aware, and that would almost certainly have occurred incrementally over a very long period time.

Just consider how when you have some sort of problem you must work through it, and then consider how long it takes to come to a viable solution for that problem. Now imagine having nothing with which to use as a springboard for your ideas. The Creator's first point of conscious could have been nothing other the realization of **I Am**, meaning "I exist". This is consistent with what it says in the Bible where in the book of Exodus, Moses went to see a burning bush that he spotted, and God said to him "'I Am Who I Am.' He said: 'Thus shalt thou say to the children of Israel: **He Who Is** hath sent me to you.'" **"I Am"** is the initial point of realization of awareness.

For the Creator to have come upon a realization of awareness is the first obvious point of self-Creation. There would have been no one and no thing to cause this. It was likely a slow process of becoming aware. If God is real, then logically there could be no other thought at the first moment of self-conception other than "**I Am**", meaning "I Exist". From that point on, the infant core of the **I Am** self-conceived awareness could then bring thought to a higher level and Create a way to add to the first core concept of **I Am**. There truly is no other explanation for the first conscious to exist, other thoughts about the initial starting point in the evolution-versus-creation debate previously discussed are all irrational thinking.

I'll go out on a limb here and say that the Creator likely does not know how the first **I Am** thought came about, other than it just occurred at some point. There would be no reckoning of time or concept of any existence before that point, and the time it took for the very first conscious to come about had infinite past to occur. I can only assume that even the Creator has no idea of how long ago the first point of conscious **I Am** occurred, other than that it was very long ago and likely occurred over a long period of existence. Hence we have a timeless Creator, although there was probably an initial moment where awareness began but it was likely vague. We have to keep in mind that there would have been no articulation of thought like this subject is being conveyed to you with as you read this. It would have simply been **I Am** and nothing more than that. Then over a long period of limitless duration of existence, other building-blocks of thought would logically have had to have been conceived. At a point so very long ago in terms of man-years at the initial point of self-Creation was the **I Am**.

Godless Tactics

In the previous section discussing the "**I Am**", in reference to the Creator, we arrived at the realization that there is no other rational logical possibility. Since there was nothing known to the

Creator before that time, and nothing afterwards that the Creator did not Conceive and then Create, it is logical that the Creator would claim superiority and extreme longevity. But this won't stop detractors from using various tactics to try to refute the concept of any sort of rational Creator existing. One of the very big problems that supporters of Creation have is that they all too often believe lies about the Creation account due to bad Bible translations, and thus they are easily mocked for their ignorance. Issues with Bible Translation are discussed in the book *Understanding The Bible - The Bible How-To Manual AND The Things We Don't See*.

Sometimes it's tough to decide which side is more ignorant. Bible supporters will say "The Bible said it, so it's true." And that may in fact be so, but not with most interpretations and translators' interpretations that most Bible thumpers use. I can hear cheering from the evolution section right now, because that was a profoundly true statement. But not to let anyone get too full of themselves, let's examine science's bible-of-evolution. With the Holy Bible, at least it is a writing of respected antiquity, but with the bible-of-evolution it is quite a different story. With your eyes wide open, if you have ever read the core works written by evolution proponents, you will then quickly see the convoluted tales of blind faith that are purely speculative woven throughout the various evolution theories. Nearly all of the evolutionary tales are speculative, but to sound more authoritative, they will take certain basics of biology and extrapolate that to full-blown end-to-end evolution in order to help make their convoluted theories look plausible.

So which side is the bigger offender with their interpretation of the data; the ones who have documents dating back several thousands of years which are supported by other world histories? Or is it those who have, only in recent centuries, compiled an ever-changing set of data adjusting their theory every time they find an error in it? You might look at science making adjustments in the evolutionary theories as a good thing, and it is; but do you see any flaw in the fact that such "science" changes? What we believe scientifically today might change tomorrow.

While it might hurt your ego if you are a Creation supporter when the idea of Creation is mocked and derided by evolutionists, it is completely justified that they do so. In fact, I would go so far as to say the Creationists often deserve it. It doesn't matter which side you are on in the debate, because if you are being irrational, then you deserve to be called out on those irrational thoughts that you are pushing onto others.

Atheists and other Creation deniers often spew a litany questions for consideration. These rhetorical questions are obviously meant to mock the past errors of Christianity's interpretation of Creation. Yet this has nothing to do with the existence of a Creator. Atheists excel in discrediting the Christians, the Bible, and the Creator because they attack, attack, attack. And as long as the Christians are backed into a corner defending themselves while receiving the severe beating in the debates, the result is that the Christians are not be able to think clearly. This usually provokes the weak comeback of "The Bible says so".–When you make a claim, you had better be able to explain it with more than "The Bible says so."

The issue of The-book-told-me-so is no different with the evolution side of the debate. The big difference between sides is more in temperament than it is content. Christians tend to be a bit gentler, and the atheists tend to be more belligerent. But here we must take a look at this a bit closer. We typically align atheists with evolution and Christians with Creation, but it's not that simple because there are many Christians who have bought into the evolution ruse, and there are even some atheists who can't find the fullness of logic in the amoeba-to-human scope of evolution. If we cannot explain everything about a theory, then we should probably shut up or preface our statement with something to the order of "I Am uncertain, but this is what I currently think..." and then go on to elaborate on our thoughts.

What everyone needs to do is to seek Truth and Fact and stop to think before the words flow out of our mouths. I have heard more than my share of foolish Creation perspectives that are neither viable nor logical, and the same is true of evolution. We must use logic and reason, and while some Christians might not want to hear this, it is more true that Christians must use logic and reason, than it is for evolutionists or atheists to do so.

Christians, by the very fact that they claim that there is a Creator who Created **everything**, makes them far more culpable in their beliefs and theories. If we are, in fact, Created and God gave us superior intellect, but we then use that intellect illogically, it seems to me that doing so is an insult to the super-logical super-intelligent Creator that the creationists claim brought all things into existence.

Evolutionists and atheists don't have any god that they feel they need to answer to, so they can say any nonsense they want to make up as long as it pays the bills or satisfies their need to slander Christians. In the end, a lot of the problem comes down to assumptions.

Incorrect Assumptions

Nearly all of the argument in the evolution-versus-Creation debate comes down to incorrect assumptions, that is to say incorrect *conclusions*. A great portion of the debate occurring in the evolution-versus-Creation controversy has nothing to do with the actual debate and is more about religion. This is true of both sides of the debate, although it is easier to detect on the Christian side of the debate because, as discussed in the books *Understanding The Church - Upon This Rock I Will Build My Church* and *Understanding The Bible - The Bible How-To Manual* AND *The Things We Don't See*, there are many people who use poorly translated late post-Reformation Bibles that they wrongly assume to be accurate because they are wrongly told that "The Bible is inerrant!" This problem is no different with the evolution gang who bows before the god of evolution and assumes that everything they read is "factual" and properly reiterated and then finally properly interpreted by them.

The assumptions made with the Bible are somewhat different in that those people are assuming that the translations of written histories are correct. So while the translations might be inadequate, they are translations of historical events that were written at or very near the time that the events occurred, with the exception of the Creation text. But the story is very different with evolution. With evolution, its very premise is built *only* upon assumptions about the "fossil record" and then even more flawed assumptions derived from actual observation of that fossil record. And finally these assumptions have to be further interpreted by the reader of the documented assumptions. This is where your ability to properly reason comes in very handy.

An Ability to Reason

A strong ability to reason is unique to humans when compared to animals. Now, there are many evolutionists who will debate that point, but it is true. The reason that this is debated is that animals clearly use reason to some extent. But with humans it is not our ability to reason that makes us unique, rather it is our level of ability to reason that is the key point that makes us unique.

Rational reasoning is one of our key attributes that has made man thrive in a way that gives us articulate superiority over the animal kingdom. The level of logic man uses is unparalleled in all of the living creatures on Earth. And only when we violate our capacity to reason do we become more animal-like. From a God perspective, we shed our strong ability to reason when we sin as we deliberately harm others. Acts of violence are not thought out well and are typically more of an impulsive act, rather than an act of pondering action and consequence. From a Biblical perspective, this is another of the things that makes man unique—we know the difference between *good* and *evil*.

There is a difference between understanding that there is a distinction to be made between good and evil, versus being able to determine good or bad. Knowing that something is bad or will harm you is displayed by animals all the time in the fight-or-flight responses, but knowing between good and evil is very different from that. When we understand that something is evil, we then specifically understand that taking action to deliberately harm an innocent person is an evil action. Without an ability to reason through good and evil, we would not be able to tell the difference between evil intent and someone being hurt by accident. The result of both is similar, but we have a uniquely strong ability to be able to know that deliberately harming others is evil.

Our level of ability to reason is not because we know the difference between good and evil, but it is one of the reasons why we are able to tell the difference. According to the Bible, Adam and Eve ate from the forbidden Tree of The Knowledge of Good and Evil, and when they did, their eyes were opened. This enhanced their ability to reason even more. Upon their Creation Adam and Eve were God-like, having been Created in the Image of God. We can debate them having God-like qualities, but it is consistent with the Bible in that we are uniquely different than the animals, and Adam and Eve could discern and Create at a level that far surpassed any other living creature.

Not only can mankind, or more properly stated "man", deliberately create little duplicates of ourselves, but we also can create all of the manmade wonders we see around us that we have pondered, invented, designed, and then built. Our ability to see in advance and plan and then Create something that never before existed, such as something that is extremely detailed and has tremendous thought incorporated into it like mechanical devices or electronic devices, is on a level that no animal has ever been shown to exhibit and likely never will. Our Creative ability is consistent with the Bible and it has to do with our superior ability to reason—an important part of our In-the-Image-of-God nature.

When God **became** at the point of "**I Am**", the ability to reason was borne. Reason was among the very first acts of anything ever done in all of existence. And then to use that awareness to further Create, shows intelligence on a level that is difficult to comprehend. If atheists are correct, then that's all bullshit. But then the atheists have many questions to answer that evolution struggles to explain. Realizing that the Creator is self-Created is a logical point of understanding and is the only realistic possibility. If anyone has another that does not violate logic by trapping itself in an endless cycle of violating reason by having to repeatedly ask, "Yes, but how did that get there?" I'm sure that everyone would like to hear it and would be very attentive to the entire explanation of such theories.

The Catholic Catechism has a statement in it that specifically says that we are not supposed to question some things, but that violates our design and what Christ told his Apostles. The books *Understanding The Bible - The Bible How-To Manual AND The Things We Don't See* and *Understanding The Church - Upon This Rock I Will Build My Church* have more information about this. The Catholic church says to "not indulge in curious inquisitiveness by investigating and scrutinizing..." But Jesus Christ said "Ask, and it shall be given you: seek, and you shall find: knock, and it shall be opened to you. For every one that asketh, receiveth: and he that seeketh, findeth: and to him that knocketh, it shall be opened." I'm going to have to go with Jesus on this one. If God is actually real, then it is in our God-like nature that we are to be inquisitive, and we must ask and seek and knock in order to know God better.

Chapter 3

Man - The Higher Creation

Can we prove that God actually exists by exploring the evolution-versus-Creation topic? I do believe that any rational mind can do so, but only if we suspend all judgment and make the assumption that a Creator truly exists and then ponder that Creator from that perspective, rather than assuming that there is no Creator as the starting point to prove that a Creator exists. In doing so you should arrive at the conclusion that the Creator is extremely logical and fair and wants only what is good for the Created creatures. And it is this general Creative nature of the Creator that has been instilled into "man".

The basic subject of this book is the arrival of "man". Man can reason through how a self-Created Creator might have come to be, but how did things get from a self-Created Creator to "man"? How did something come from nothing? We can logically suggest a self-Created Creator because conscious is not made from any substance or thing–it just exists. Once conscious began to exist with **I Am**, it is just there and is non-scientifically-tangible in its existence. But how do we get from a non-tangible conscious, all the way to tangible matter? (Some of this is discussed in greater detail in *The Science of God Volume 1 - The First Four Days*.) Our collective scientific quest to understand origins has led us down a trail of particles.

We first see an object and then we wonder what that object is made of. As we study the object, we find that it is made of particles or cells. Then we look deeper and find that those are made of molecules, and we look at the molecules and find that they are made of the base elements that we find in the *Periodic Table of Elements* listing the various known atoms. To go further, we then break those atoms to pieces to find protons and neutrons and electrons, and then even deeper within those we find quarks. Then beyond quarks we theorize that things are made of "strings" or "vibrations". However, at this point the items are so small that we have no means by which to prove our theories or their existence. So what is it that is potentially "vibrating"? Since nothing exists, these "vibrations" could not be vibrating anything.

There is a point where we have to admit that everything is made from nothing. Since there was nothing before something existed, it means there was nothing to make the something with. Weird? Yes, **very** weird!

But logic demands us to admit that there is no other option, because if we say that there was something there that was used to make anything, no matter how small or insignificant it may have been, we then have to ask where the original something came from, and now we are caught in the endless *Yes-but-how-did-it-get-there* trap. This brings us back to the concept of conscious. Biblically speaking, we might refer to this as "spirit". If you look up the roots of the word "spirit" you will find that it means *breath*, or *breathe*, or *blow*, or *wind*. All of which alludes to something that is not as readily tangible and is invisible to the naked eye. Scientifically, we know that "breath" has substance because we breathe the air and air is both scientifically detectable and physically touchable. Yet the air is really what we are detecting rather than the actual breath.

Even when discussing big bang theology, we have to admit that everything came from nothing. And in this case it is irrelevant as to whether all matter was in a point so infinitely small so as to be scientifically undetectable, or simply did not exist until the ignition of the bang. Either way there could not have been anything there before there was something there, or we are then trapped in that pesky infinite *Yes-but-how-did-it-get-there* cycle.

Which of these is more logical and viable: Nothing spontaneously banged into everything, or the conscious **I Am** came to be, which needs no substance to exist?

Higher Mind

When analyzing the ideas of *conscious* and *thought*, we quickly come to the question of what those are or what they are made of. But this goes back to that initial **I Am** where that initial point of conscious came about without intervention or matter. If we go down this line of thinking, we are forced to realize that conscious would have to develop additional systems of thought before anything else could occur. If you consider *thought*, there are not a whole lot of needs, rather there are just a relative few concepts that need to be developed and then all other new thought can be built upon those building blocks.

This might seem odd or a bit grasping at first glance, but consider this: We can take switches that are either on or off, thus having only two fundamental states or positions. Then we can take several of those switches and assign values to each of them. With these values, we can make cumulative values. So, big deal, it's just a bunch of switches, right? Yes, but that is the basis for all computers. Computers use a switch system of **on** or **off** and those switches have a value doubling with each switch. Computers basically use sets of eight switches. The first is valued at 1, the next is 2, then next is 4, the next is 8, the next is 16, the next is 32, the next is 64, the next is 128. This 8-switch set offers a total possible value of up to 255 when all added together, with an additional 256th all-switches-off state value of zero. And everything done with computers is done with this basic 8-switch-set model. The Creation of computer logic that mankind has built is diverse and logical. Is there a better way? Possibly. But the 8-switch-set model is very versatile and we have not yet found anything in the computer realm that we cannot accomplish with it.

Now, let's get back to the **I Am** conscious. There are likely more than two initial levels or states of thought. There are probably a small handful of basics from which all other thoughts proceed. So here is the question at hand: We can logically accept that the **I Am** conscious could come to be out of vast nothingness. That's easy to accept, even scientifically. But is there a way that this intangible **I Am** conscious could cause tangible matter of any sort to come into existence?

When we consider string-theory where we speculate "strings" make up quarks, we think in terms of "vibration". Our minds always want to gravitate to what we know and have experienced, so we understand vibrations as something tangible. For instance, when a piano string is emitting sound we can feel the strings vibrating. But vibration in that way could not occur if nothing is there. So, if we have no air and no substance, then we have nothing to vibrate and no transmission of vibration. If we have nothing then nothing can vibrate, although light does have "waves" but needs no known tangible substance to carry those waves. So the initial substance including light would have to come from *thought* itself.

Since the **I Am** conscious had eternal past to initially develop, and had that same eternal past to develop the base aspects of thought, there is no knowing all of the various unsuccessful attempts that might have been made in order to arrive at how we think today. It is likely that most of the base aspects of thought are logically established and that others were potentially considered, but had no logical value and thus were not sustainable, meaning that they would not persist and would essentially go away.

When going down this line of thinking, we at some point have to realize that the *matter* substance in Creation is Created with *thought alone* and not from any substance. This means that thought is supreme and the **I Am** conscious would be very ancient and would have experience-unparalleled and wisdom-beyond-compare. In *The Science of God Volume 1 - The First Four Days*, substance is referred to as "pre-matter", which is interestingly referred to in the first sentence of the Bible as "heaven and earth" where it says, "In the beginning God Created heaven and earth. And the earth was void and empty." This is quite interesting phrasing, because at this point we have some sort of substance or pre-matter that is described as only two states, that is to say "heaven" and "earth". With the "heaven" pre-matter substance being a higher state and the "earth" substance being a lower state. To understand this in more detail, consider reading *The Science of God Volume 1 - The First Four Days*.

Then Genesis says: "And the earth was void and empty." Here we have some sort of "earth" pre-matter substance that is "void and empty". This indicates that it exists but it is next to nothingness. So from the Bible's text we can easily suggest that thought Created "heaven" pre-matter substance and "earth" pre-matter substance and that all things come from those two pre-matter substances in a somewhat similar way that computer switches have a higher and lower position or state.

Next is the point where higher thought and higher mind was invoked, "and darkness was upon the face of the deep; and the spirit of God moved over the waters. And God said: Be light made. And light was made." The question here is, was light a third type of substance or it is some sort of higher function built upon the "heaven" and "earth" pre-matter? The rest of the first few days of Creation are discussed in the following books:

1. *The Science Of God Volume 1 - The First Four Days*

2. The Science Of God Volume 2 - Day Three - Gravity, Land, Seas, and Evolution of Plants
3. The Science Of God Volume 3 - Day Five and Day Six - The Creatures - Revolution or Evolution

Human Ingenuity

We will be jumping forward and backward in time as we try to understand man's arrival and whether we were Created by the conscious **I Am**, or if we evolved. Now we're going to skip ahead to day six when man was allegedly made. Let's just assume that God exists and Created Adam and Eve as it is stated in the Bible's Genesis One.

Early evolutionists speak of the differences between the lowest man's mind and the highest man's mind, that is to say "aboriginal people" versus "refined people of culture", believing one to be more "evolved" than the other. This line of thinking is built on racial prejudice.

I submit that any newborn baby from any group anywhere in the world at any time in history who is not mentally impaired due to any sort of environmental damage or abuse, can be taught to be a brilliant "civilized" scientist, or engineer, or musician, or chef, etc. when placed into a proper nurturing environment with resources from which to learn.

Man has proven time and time again to display a tremendous amount of ingenuity and Creativity in a way that no animal has ever been shown to do. What we have to ask is, where does our unique ingenious nature come from? Did our mind evolve to a point of having a great capacity to Create? Or were we Created with this unique ability ready-to-go? Evolution would have us believe that we slowly, over time, developed our intellect through evolutionary progression. But if this is so, we then have to question why there are no other creatures that share a similar advanced nature as man. This might appear to be a silly point on the surface when viewing this from the evolution perspective, but consider this: There are hundreds of basic creature types and most of them function in much that same way with varying intellect or ability to reason. Some people will study chickens or maybe dolphins or moneys, with each having differing cognitive abilities, but man excels so far outside of all animals that it defies the evolutionary model.

Since amoeba-to-man evolution is believed to be progressive and the key elements of the successive chain are allegedly still intact, why are there no other creatures that are anywhere near to where man is intellectually, creatively, and articulately? When using the evolution model, we would expect that there would be other creatures similar to man in that they could think and reason to a level similar to man, but would be very different in form.

For instance, elephants are believed to be "smart" as are dolphins, yet they greatly differ in form. There are many such examples regarding animals, but mankind has no such counterpart of like intelligence regardless of form. One could argue that the different races are such, but the form is nearly identical where with elephants and dolphins they are very different. So, if we found some dolphin-like creature that was building cities out of dolphin-made materials and communicating through some means equivalent of a telephone system then we would have to admit that evolution is undeniable. But since, to our knowledge, no such advanced creatures exist, man stands alone in this advanced nature.

We could say that monkeys are the counterpart to humans, but that is more like comparing a dog to a worm, the difference in intellect between monkeys and humans is vast. Humans have no evolutionary equivalent in any branch of the evolutionary charts. There are no articulating fish or birds or beasts. Humans stand apart from all other creatures with our human ingenuity. This is consistent with the Bible where it says "And God Created man to his own image: to the image of God he Created him: male and female he Created them". The **I Am** conscious made us in the uniquely

Creative image of **I Am**. This reconciles with logic, but a missing counterpart for a creature of superior intellect in the evolutionary chain does not reconcile with even evolutionary logic. The progressive chain of evolution demands a counterpart, or at least a progressive chain of intellect close to man, but no creature exists anywhere near the intellect of "man".

All Kinds of Things

When we close one eye and examine evolution, we can easily surmise that we evolved, but to do so we must ignore half of the logical analysis. We can look to simpler lifeforms that currently exist during our own time while we are alive, and then examine them from one to the next in order to create a very compelling trail of progression leading all the way to man. But as mentioned in a previous section, this ignores the peculiar absence of similar intellect in other branches of the evolutionary tree. All creatures currently have existing counterparts with similar intellect and instinct in other branches of the evolutionary tree, **all** except for man.

Of all of the different kinds of creatures, man stands alone. No other creature drills for oil and mines for iron etc, and then takes that material and refines it into pure material and from that pure material makes items like automobiles or any other such device. Man is uniquely creative amongst all creatures, which is consistent with the Bible where is says that we are made in the image of the Creator—that is to say "God" who is the self-Created **I Am** conscious in whose image we are made.

The Bible's Day Six text speaks of "kinds". Here again is the text for your convenience "And God said: Let the earth bring forth the living creature in its *kind*, cattle and creeping things, and beasts of the earth, according to their *kinds*. And it was so done. And God made the beasts of the earth according to their *kinds*, and cattle, and every thing that creepeth on the earth after its *kind*. And God saw that it was good. And he said: Let us make man to our image and likeness: and let him have dominion over the fishes of the sea, and the fowls of the air, and the beasts, and the whole earth, and every creeping creature that moveth upon the earth. And God Created man to his own image: to the image of God he Created him: male and female he Created them. And God blessed them, saying: Increase and multiply, and fill the earth." Take note of the term "kinds" that I have italicized and also notice that with man there is no vague "kind" listed, we are simply "man". But even if we had been listed as a "kind", it is specifically man-kind. There is no other creature on Earth that can be included in the "man" category. Terms like "beasts of the earth" are very broad and can include lions and giraffes and squirrels etc. Terms like "and cattle" are also very broad and can include cows and horses and donkeys etc. And take notice of "and creeping things", how broad is that description? If I ever need a patent attorney, I want it to be God, for the sheer breadth of scope covered by the patent terms. "Man" is clearly different as is stated in the Bible. Sure, "man" could fall under the "beast" label, but, Biblically speaking, it is important to note that "man" has its own unique Creative point of entry in the text.

In the evolution-versus-Creation debate, Creation supporters consistently trap themselves by allowing the evolution believers to set the terms of definition for "kinds" and for "species". This tactic has worked out very well for those who believe in evolution because it corners the Creation people in an unescapable moving target of terminology. This irrational tactic is unfair to the blind-faithed believers of Creation because it inserts narrowness of scope into the Bible's Creation text that is not actually written or implied in the text. The list of "kinds" in the Genesis Day-Six account is extremely brief and broad, "Let the earth bring forth the living creature in its kind, cattle and creeping things, and beasts of the earth, according to their kinds." So any talk of cat or dog or any other specially defined creature is putting limits into the text where no such limits are defined. The only Day-Six limits here are "the **living creature** in its kind, **cattle** and **creeping things,** and **beasts of the earth**". This vague list is so very broad and is able to include any existing creature that fits in the very broad categories stated on days Five and Six. So there is no need for any Creation supporter to explain the arrival of cat, or dog, or any other animal in this regard.

One area that is difficult to get a handle on when looking at the evolution side of this topic is the parameters for the definition of a "species". In the Bible where we see the term "kind", the Latin Bible version uses the Latin term "species", which is where the scientific usage of the term species is originally derived from. But our scientific hairsplitting examination of creatures very loosely defines species; and those parameters can and do greatly vary from person to person and from scientific discipline to scientific discipline. This makes sense since there is obvious progression of evolutionary adaption, because at some point the people defining the differences of one bird to the next bird must make a distinction and declare a new "species", but this is a very arbitrary aspect of science. Thus, there is no end to species or kinds as both terms are broad and are defined on a per scientist basis.

Mankind

So what is a "kind"? A "kind" in Genesis is "each after their own kind" but does not make any distinction other than what is stated in the Bible. Someone could argue that there is only small list of "kinds", yet today we see many more kinds/species than are listed. This is true, but again we then get into the issue of definition of "species". The broadly stated "kinds" can, by our modern description, contain a vast variety of creature forms. This is very different with man because what we see in the world today and throughout all of archeological history is that man has remained man since man appeared in the buried archeological vault we call "Earth".

All of our archeological evidence supports the Bible's account of "kinds/species". Evolutionists get this wrong because the deceptive illusion is so strong that they will force in false data where no data exists. This is done in order to fill in the spots that they interpret as "missing" data. Meaning that evolutionists *expect* to find transitional forms between primates and humans, and so they vigorously seek to find those forms. Any evidence found that has any deviation from typical human or primate form will be placed in the gap unless the deviation does not meet with their progressionary evolution needs. Over the years there have been many fraudulent findings where people wanted the recognition and money, so they would combine parts from different creatures or lie about location or depth of dig or any other clever deception to further their lust for fame and fortune. What is worse is that some of these early deceptions are still taught in modern times as "fact". But let us not fool ourselves, while these things have and still do occur, it is certainly not true of all evidence that has been found.

Made from Slime

Genesis One is sparse on details regarding methods of Creation other than to make broad-scoped statements that all allow for a great deal of versatility. There is little mentioned about Adam being Created or the methods used to do so.

However, in the English version of the Douay-Rheims Bible in Genesis Chapter Two it says "And the Lord God formed man of the slime of the earth: and breathed into his face the breath of life, and man became a living soul." And also, in the apocryphal book *Tobias*, or *Tobit* as it is often called, found in many Bibles, chapter Eight verse Eight it says "Thou made Adam of the slime of the earth, and gave him Eve for a helper." So this, is saying that Adam was made from the "slime of the earth". In the original Latin translation, it uses the word "limo" which can be translated as "slime" or "silt". This notion of being made from "slime" is really quite important if we are going to try to logically reconcile Creation with actual science, rather than taking the hocus-pocus-POOF!-and-suddenly-Adam-was-there approach. There are two key points to note in the Creation text; First is that Adam was made from "slime of the earth", and second is that God "breathed into his face the breath of life".

Did Adam Have a Belly Button?

For space sake I don't want to elaborate on the details of how the creatures were formed, to read more than what is mentioned in this section read *The Science Of God Volume 3 - Day Five and Day Six – The creatures - Revolution or Evolution*. The three previous volumes of *The Science of God* are divided into three key topics

1. *The Science Of God Volume 1 - The First Four Days* is about the physics and astrophysics aspects of Creation.

2. *The Science Of God Volume 2 - Day Three - Gravity, Land, Seas, and Evolution of Plants* is about gravity and some other astrophysics points along with the very important arrival of plants.

3. *The Science Of God Volume 3 - Day Five and Day Six - The Creatures - Revolution or Evolution* is about the arrival of the animals including the Day Six animals.

4. And then there is *The Science Of God Volume 4 - Day Six - Evolution versus Man - In Our Image*, which is this book that you are reading now about the arrival of Man.

5. In addition to those 4 books there is also *The Science Of God Volume 5 - Boats, Floods, and Noah - The Deluge* discussing the details required for the possibility of the Biblical global flood to have occurred, along with the connections that all has to the evolution-versus-Creation topic.

In brief, *The Science Of God Volume 3 - Day Five and Day Six - The Creatures - Revolution or Evolution* explains the needs for animal-life to thrive. The events that are discussed in *The Science Of God Volume 2 - Day Three - Gravity, Land, Seas, and Evolution of Plants* are regarding the arrival of plant-life that is required in order for the creatures from Days Five and Six to be able to live. But what we want to know is how they got there. With animals it is unlikely that God did POOF-like magic to conjure up animals. The text says "Let the *waters* bring forth" for Day Five, and "Let the *earth* bring forth" for Day Six. But for Adam it does not give the "water" or the "earth" the power to specifically bring forth Adam. Adam was uniquely tended to according to the Day Six text.

If you read *The Science Of God Volume 3 - Day Five and Day Six - The creatures - Revolution or Evolution*, you will recall the profound dilemma of "What came first, the chicken or the egg?" It is unlikely that the Creator just suddenly POOF'd things into existence. If we were to witness the spontaneous arrival of a creature suddenly appearing today, then I could take that thought seriously. However, since that is unlikely, then it is highly probable that the creatures had to gestate to a point of relative maturity. And here is where things get interesting; this means that the chicken almost certainly did not come first. Thus we are left with the egg. So did God make a bunch of eggs that hatched, thus delivering all sorts of life? That's not likely either. So then, how? How could this have occurred without the chicken or the egg?

To really dig deep into this, we have to examine the simplicity of a smaller bird's eggs and take note of the length of required incubation period for that kind of bird. Many small-bird eggs will incubate in as little as about ten days before they begin to hatch. This brief incubation window combined with a rapid reproductive cycle of less than three months for some birds allows for four generations of those birds inside of a single year's time. This prolific nature would produce countless birds in just a few years from only a single initial pair.

But whether egg-borne or mammal, we still have to get to a place with no egg and no initial parent creature. This is where the egg becomes so interesting. Eggs are relatively simple items and all laid eggs share a similar makeup having a yolk or fat part and the albumen or protein part. There is a bit more to it than that, but that should give you an idea of the simplicity of materials needed for life to be Created. An egg is little more than an egg until it gets a bit of extra information from a fertilization event. Once the fertilization event occurs, then within twenty-four hours there is noticeable activity on the yolk. So the primary needs are, fat, protein, and, finally, information.

Since we can, on an evolutionary basis, imagine that a lightning bolt hit some raw materials to initiate the chemical beginnings of an amoeba, it should be no stretch that a similar event could occur in a similar slurry of material, thus initiating fertilization. The only difference here with this approach and the evolution approach is that evolution is believed to, slowly over millions of years, have added information and then diversified to form all of the creatures that we see today, versus what is being stated here where the lightning bolt contained basic information that it then infused into the slurry of required materials. This should not be a stretch for any evolutionist to accept when considering the technical leap required to arrive at the theory of lightning striking chemicals, thus causing amino acids to develop and then evolve over very long periods of time.

The difference in the standard evolution theory and the theory being proposed here is that in this theory there is variance in the slurry that was available throughout the earth, and those varying materials would result in a variety of differences of outcome. In the theory being presented here, the slurry's composition determines the variance in the various birds, for instance. This is somewhat similar as is suggested in evolution where the environment influences the outcome. But evolution suggests that it happens over extremely long periods of time, where the details presented here state that the environment influences were immediate and offered great versatility of fully formed creatures in the very first generation.

Now, since this is discussed in more detail in the aforementioned volumes of *The Science of God*, we don't want to spend much time on this aspect here (See the first 3 volumes of *The Science of God* for more detail.), But in our modern world, we communicate through electricity every moment of every day, so sending data via electrical impulse is provably and undeniably possible and is done on a constant basis today.

The information from electrical lightning strikes only needs to have basic design parameters that are shared by all similar kind creatures. The rest of the influence forming each creature-kind came from the immediate surrounding materials and climate conditions.

When discussing eggs, the infusion of information is very plausible, but what about mammals that are attached to the mother by an umbilical cord? How would they be able to gestate? To answer, this we first have to ask: Do birds have belly buttons? Here you need to understand that birds **do** have belly buttons where they were attached to the yolk-sac, which is how they drew the needed nutrients from the surrounding materials. And it is not only birds that come from eggs.

Discussing eggs does little to help us understand the arrival of mammals, but it has been proven that an egg can be incubated without being inside of a shell. This makes it very close to being nothing more than a puddle of required materials needing only a spark of additional information to be infused into that puddle. The same can be said of mammals where they only need the basic materials similar to the needs of an egg plus a little extra information, and they can be brought forth in the typical gestation period of each particular "kind".

If Adam was Created with a POOF!, then the likelihood of him having a belly button is about zero. However, if Adam was made in a manner similar to the other creatures suggested here, where he would grow out of earth-slime that was infused with information via a lightning strike, then the probability that he had a belly button is about one hundred percent.

Since the **I Am** Creator has no tangible hands with which to form Adam, the idea of thought-influence is our only known remaining option to cause Adam to be Created. Now, since we have undeniable conclusive proof that information data can be, and is, sent via electricity, and since we are quite scientifically confident that our DNA is information data instruction, we can then assert that the **I Am** Creator could cause thought information data to flow to the slime by means of electrical impulse via lightning.

When we accept thought data being carried by lightning and infused into the slime as a plausible solution, the question then becomes, how long would Adam have been in the slurry of slime? Since Adam and Eve's ages are not specified in the Bible, we really don't know how old they would have been upon coming aware. We tend to picture them as twenty-something adults, but it is unlikely that they were the equivalent of that in actual growth and maturity. It is highly possible, if this idea is correct, that Adam was in the slime for quite a few years until he would have been physically capable of moving as a child of maybe five or six years old or maybe to the age of early post-puberty. But that is complete speculation because there simply is no mention of age. So, yes, Adam had a belly button.

This approach to the arrival of the creatures and of man does not require evolution, but it uses the same initial electrical point of life. In the view being suggested in this book, the information for the basic kinds and for Adam would have been instantly infused into the raw-material-slurry by lightning strikes adding the critical thought information sent by the **I Am** Creator, rather than eons of subsequent generations each adding tiny bits of information as is boldly stated in the evolution approach. It is likely that there was a diverse array of creatures within each kind for the initial generation of each creature type within that "kind" for what we today call "species". But as for Adam, Genesis states only one.

"Man" has no broader scope stated in Genesis. If today we had other man-like creatures then we could consider the scope to be much wider, but that is not the case with man. Adam is unique in Creation in this respect, and our current day evidence matches the Biblical narrow design scope for "man".

It is interesting to note that in Hebrew the name "Adam" signifies "son of the red Earth" and "Adamah" is ground or earth. The likelihood that he was actually called "Adamah" is very high. But due to translation issues and the fact that "h" and "ah" are sometimes barely pronounced or are silent, those letters or sounds may have inadvertently gotten left behind in translation. We will likely never know for sure, but within the name "Adamah" is a great deal of other meaning that is too lengthy to put into this book. Adamah is indeed conceived of "slime of the earth" by the very essence and meaning of that name.

Chapter 4

A Way to Perfection

As discussed in *The Science Of God Volume 3 - Day Five and Day Six - The creatures - Revolution or Evolution*, the probability of some sort of adaptive evolution most likely occurred in the other creatures over the many years since the initial Creation events occurred, but with Adamah we do not see any deviation. I understand that there is an army of evolutionists that will argue this, but there are **no** transitional forms leading to man that cannot be either clearly placed squarely in the "primate" category or clearly placed in the "man" category without having to distort reality to a considerable degree. Setting aside any thought of evolution for the moment and analyzing only the Genesis One text, we have to realize that "man" was the last of Creation. Based upon our nature, when we create things we typically seek to achieve perfection of our creation as we conclude our creative process.

It is common for creative people to make those finishing touches to their work. Where is this common obsession from? In the context of Biblical Creation, I would have to attribute it to being Created "in the image of God". Since man was the last of Creation and was made "in the image of God", we really have no choice but to view that as—The finishing touch!

Evolution Absurdity

In the first three chapters it was noted that, logically speaking, everything that exists had to ultimately have come from nothing. I assume that rational evolutionists would concur on that particular point. However, the idea of evolution comes long after the initial **I Am** aspects of existence. In fact, if we are only focused on evolution, then none of that stuff matters. So let's forget about the initial conscious etc. and begin from an Earth of chemicals and slurry and lightning.

We can assume that there is likely more than one evolution initiation theory, but here we will stick with the most prominent. Evolution's primary theory of lightning striking water rich in minerals, thus forming the amino acids that are the building blocks of life does at first glance appear to be a plausible theory. But it fails to explain how things progressed from that point onward. And further, depending upon who you discuss this topic with, it is believed by many that this was a rarity and a bit of a fluke, yet from it a small amount of amino acids were formed and from that *all* life allegedly eventually came. This perspective makes the odds of life ever occurring at all extremely low, even if only one single primitive cell. The idea that this would have been a rare fluke event seems to me to be unrealistic. That idea is simply absurd. If this occurred at all, it would have been occurring all around the world.

If evolution has any integrity whatsoever, it must dispense with these rare odds scenarios. The difference between the Creation concept being conveyed in this book, and that of evolution, is that with the concept in this book it is the lightning that is passing information given by the Creator, where with the evolution theory, on the other hand, there is no information being passed at all, but rather only an electrical jolt.

When we allow information to be added through lightning and lightning to be occurring abundantly nearly everywhere, we then can easily arrive at life much more quickly, fully, and abundantly. All of which confirms what we actually see in nature **and** in the Bible.

Since amino acids are, in fact, a building block for life, and without information, those amino acids will basically do nothing, the odds of any life occurring is zero in the case of evolution. The base materials of DNA need *order* to be useful, and order is information. So, when we allow information to be infused into the slurry of materials, we then have life! And when we dispense with this evolutionary rare fluke nonsense, then we can have life abundantly, life which is not vulnerable to any disruption that would likely have stopped the alleged initial single point of amino acid formation that is suggested in the prominent evolution theories.

Can We Be Too Precise?

If we think through this in a logical way, we can take many actual scientific discoveries and merge them with the Bible without bending any scientific rules or any of the laws of physics. This allows for the "earth" to bring forth the creatures after their kind with many variations of each kind arising in abundance all within the same time proximity. And then from there forward, slight deviation could occur creating ever more diversity as time progresses, but only within the parameters of each initial "kind".

When it comes to defining species, evolution struggles a great deal, but Creation supporters struggle even more. As mentioned early in this book, the evolutionists tend to inadvertently, but also often deliberately, back the Creation people in to the species corner for very public floggings. This occurs because the

Creation side foolishly allows the evolutionist to add words and thoughts that simply are not there into the Genesis Creation text. This successful tactic is done by increasing the species diversity with every breath taken during the debates. What the Creation supporters are missing is that defining species is not a precise practice. To make matters even worse, Creation supporters, when pressed on the issue, tend to take a hard stance by not allowing *any* change to occur at all, other than minor adaptive changes. This traps those supporting Creation because they have blindly taken the evolutionist's bait.

But let's turn things around a bit. When the term "species" is used in an unrestrained manner, then that debate target moves as it rolls over the toes of the Creation crowd. But when we tighten the parameters of scientifically defining species, and then further look at the actual Genesis One text, we must realize that "each after their own kind" is a very broad statement that has only one restriction which simply is "each after their own kind".

As explained in *The Science Of God Volume 3 - Day Five and Day Six - The creatures - Revolution or Evolution*, a "kind", or in Latin a "species", has the implication that when two pigeons breed they will have a baby pigeon. They will not produce a chicken or a pheasant, or a monkey for that matter. Their hatchlings are going to be after their "kind", thus they will be pigeons. There are no such restrictions that would stop minor deviation from generation to generation to eventually form a variation of those pigeons, even possibly to a point of not being recognizable as a modern "pigeon", but rather as a unique bird of its own "species". However, as per the Bible's Creation text, each offspring will still be after the "kind" of its parents. This is undeniably true throughout all creatures, including "man". When creatures have offspring then that offspring is of the same "kind" as the parent creatures are—the offspring shares the very same form and look. There is no rational person on Earth that can deny that this occurs. Elephants do not spontaneously have alligator offspring etc.

More Than Two of Each

In a previous section discussing the slurry of slime with the needed chemicals being struck by lightning, we didn't really get into the scale of this idea. The evolution theory and the Creation position being proposed in this book both claim lightning as being a factor in the initiation of animated life. But the evolution proponents see this event is a *rare* fluke and from it all things evolved. Yet regarding it being a rarity, if we can reproduce this in a laboratory, then it would likely have easily occurred all over the world.

On the Creation side of this thought, we would expect that there would be nutrient-rich slurry conditions all around the Earth and we would also expect that lightning would be occurring randomly all around the Earth. And in that lightning we would expect information that would be transferred to and infused into the various slurry types of water and earth. This means that we would expect not only one or two of each creature at the initial Creation of the creatures, but rather there would be an abundance of each kind and likely a very diverse array within each "kind" group stated in Genesis.

If life was so fragile as evolution and typical Creation perspectives tend to make life out to be, where it "almost did not occur", then we can barely assume that we would even be here today, because life would still have the same vulnerabilities. Yet, this is *not* what we see in reality. We see exactly what Genesis says, each after their own kind, with man being uniquely separate. So based upon the Bible's Creation text, it is unlikely that the initial bringing forth of creatures was from isolated occurrences of only two of each as many Creationists and evolutionist imagine that the Bible indicates. It is far more likely that this would have occurred all around the globe causing great diversity and abundance of each kind to quickly spring up everywhere in both egg-borne creatures and non-egg-borne creatures. This is the only way to robustness and perfection of each kind that we witness today.

Chapter 5

The Breath of Life

Whether or not either side of the debate will consider the controversial views put forth in this book is unknown, but we should at least expect that neither side can tear it down like both other perspectives can so easily be torn to shreds. So for the sake of continuing the line of thought presented in this book, we will now evaluate Adamah's, or Adam's, becoming.

Not Believing Anything

Over the years in discussing the various Creation topics, I have heard on numerous occasions where evolution proponents will say that they "only accept facts and don't just believe things other people say". But of course this is absurd, because if you read a book and see the pictures and accept their account as "evidence", then you are believing what the author of that book said. So unless you're actually out there digging in archeological digs or actually studying biology by doing microscopic biology experiments, you are typically believing what you have heard or read–This is an inescapable truth. But it is not so much the information we take in, as much as it is *our proper evaluation* of that information. In other words, we can read and just believe, or we can read and reason through what we have read and draw our own conclusions based upon our own ability to logic our way through the information that we have just read. However, regardless of what we conclude, there is only one truth and it is our quest to find that truth. So even when our conclusions feel logical within our own heart and mind, those conclusions could still be in error. We all need to keep our minds open, even after we have drawn our initial conclusions, because new information could present itself at any moment.

It really doesn't matter which side is talking, because both sides of this debate have *beliefs*. Sadly, most people on either side do not use sound logic or common sense when analyzing the information presented in their own theories, let alone their opposition's theories.

Could Adamah/Adam have gestated in slime for several years after nutrient laden slime was infused with Creator-given instructional information from a lightning strike? That seems far more plausible than those POOF-instantly-formed-man Creation accounts. It seems to me that we need to evaluate the environment around us and pay attention to how things *actually* work. Allowing normal gestation periods and intelligible basic information to be infused into nutrient laden earth-slime truly explains a great deal in this regard. It supports the idea of lightning striking thus causing amino acids and it supports decisive Creation without violating either theory. This is not to pander to either side; rather it is the logical likely reality of the concept. At this point in human history, we really have no other good explanation of how the Creation of the creatures occurred.

At First Breath

For the sake of further exploration, we will accept all of the points made so far in this book regarding the arrival of the creatures and the arrival of man and move to the next problem. Genesis One indicates that the creatures would reproduce after their "kind", but with Adam the text doesn't offer the same broad scope, it simply says "man". This could allow for a larger scope of diversity of "man", but since we do not see great diversity in man we will assume that man was intended with the very narrow scope that man actually is today.

Backing up a bit to the conscious **I Am**, the idea of spirit or "breath" was mentioned earlier, indicating that something was not visible or touchable, thus spirit is undetectable by any scientific means. We breathe **air**, so the idea of "breath" is not what we tend to think of it as in this context of Creation. The idea of spirit is more along the lines of the **I Am** awareness. That is to say that the first "breath" of the **I Am** occurred very early on, shortly after the Creator's initial awareness was established countless eons ago.

At this point, this assumes that the Creator was self-Created pure thought, meaning that nothing triggered the initiation of it, but rather it slowly came about to that first point of **I Am** existence. From that point onward, the **I Am** had endless time to develop other fundamental concepts of thought. Understanding how thought works and how everything we see around us works and how the nature of man ponders the fact that we are Created in the image of the **I Am**, it is logical to conclude that a great deal of trying occurred by the **I Am** while developing the fullness of thought. Having endless time to have done so, many avenues of thought could have been explored, thus giving the **I Am** so much experience that no Created creature will ever be able to compare to the **I Am**. This would be similar to comparing the knowledge of a wise and alert very experienced one-hundred-year-old person to a one-second-old newborn infant, except with the **I Am** it would be an exponentially greater information gap compared to us as we are today.

Adam's First Breath

Following the Creation parameters of the Bible, Adamah would have likely been in the slime for a quite a while in order for him to develop fully enough to be a viable "man" that could live outside of the nutritious embryonic earth-slime. In our context here in this book, "man" is not some fully formed "legal aged" adult; rather, man is a person that could have gown to be of any self-sustainable age before having become aware.

Now, we have a diversity of creatures established with one additional unique creature that was allegedly Created in the "image" of God. Did this "man"-type creature spring to life of his own accord after having gestated long enough to be capable to survive in the world? There is no indication that the other creatures had any such special treatment. The only indication is that both the "water" and the "earth" "brought forth" all of the creatures, but with man the details are somewhat different. In Genesis Two it says: "And the Lord God formed man of the slime of the earth: and breathed into his face the breath of life, and man became a living soul." This is not so with the other creatures, at least it is not stated as such in the text. The verse about breathing into the "man" the "breath of life" is a critical part of being made "to the image of God". I suspect that if God had not breathed the "breath of Life" into Adamah that he would have been little different in behavior than all of the other animals, though unique in design. The design of the animals appears to have been flexible, allowing for vast diversity immediately in the initial generation, as well as vast diversity to develop through subsequent generations, but with man we see no such diversity.

Truth be told, based upon the way many bird species are categorized in science, humans would be broken in to several "species" groups running along the lines of race as we see man today, and this would be the case if science was actually consistent in defining species. The idea of multiple species of man is based on the moving target of defining species, which is anything but an exact science. In any case, man is very specifically more exclusive than the animals regarding divergent adaptive evolution. In other words, there simply is no evidence of such.

Did Adamah's first breath start like the other creatures? Or was his first breath the act of receiving the spirit/breath of God? According to the text, it appears that it is the latter. "And the Lord God formed man of the slime of the earth: and breathed into his face the breath of life, and man became a living soul." Now, we could write into this that after Adamah was fully formed and ready to come out of the earth-slime that God waited until then to "breathed into his face". But there is no indication of a time frame in that regard. So we could assume that God "breathed into his face" very early on after his having been conceived in the embryonic slime. The "breath" that was "breathed" into Adamah may have been nothing other than the very essence of the **I Am** who is the self-Created pattern of thought. This is likely the point where Adam's initial intellect was added.

There is nothing that has ever been scientifically shown that differentiates us from the animals, even though people have been seeking answers to this for a very long time. Some people claim that we should not question this, but with that I vehemently disagree. However, neither should we make foolish assumptions. Man having received the "breath" of the **I Am** and being separately "Created in the image of God" seem to be the only differentiating factors between us and the animals, which explains why "man" is so very different than the animals. But evolution science has yet to offer any viable answers as to why or even how man is so very different than any of the animals.

Chapter 6

Woman - The Perfect Companion

After God Breathed the Breath of Life in to Adamah, a few other events occurred, and then God took note of the fact that Adamah was alone. God then decided to Create a companion for Adamah. All of the animals already had their male or female counterparts, but not so with "man". God needed to make a perfect companion for Adamah. Now, since the earth-slime method of bringing forth appears to have offered a vast and diverse variety of creatures within each kind, it is possible that God wanted Adamah's companion to be made in the image of Adamah. The **I Am** is a unique being with no other known similar being to exist, and since the **I Am** is the only one, it stands to reason that, unlike in the case of the animals, God only made one version of Adamah. But how could God the **I Am** make an identical creature like man for man?

A Pristine World

There are many people who make God out to be some make-believe hocus-pocus POOF-like God who waves a magic wand

and then, POOF!, there it is, pristine and fully formed! And it's all lollypops and rainbows and unicorns. But that is highly unlikely. I'm not sure if all creatures have "souls", but there is a cycle of life that according to everything that we have **ever** witnessed on Earth does not deviate. This means that "death" did occur before the "fall of man". It has to have for life as we know it to exist. It is possible that the creatures were not carnivorous before that time because it says: "And God said: Behold I have given you every herb bearing seed upon the earth, and all trees that have in themselves seed of their own kind, to be your meat: And to all beasts of the earth, and to every fowl of the air, and to all that move upon the earth, and wherein there is life, that they may have to feed upon". However, based upon the Genesis One text, there is a lack of detailed information here regarding some of the "kinds". So, to insist that something is one way or another stretches the text to something that is not stated and is not even indirectly implied.

The earth would have been pristine to the extent that it was not yet deliberately altered by man, but it was likely using all of the same functions that occur in nature today, and thus would have been little different than what we experience when walking in a forest today. There is one exception here though; because while there was likely an abundance of variation and likely many of each of the variations of creatures at the inception of the creatures, there was probably a relative few in number in comparison to the multitude of variations of birds etc. alive today. So, the environment would have been very pure and largely undisturbed from a human activity standpoint.

Since there is no timeline stated other than the "days" that most people infer from the usage of that term, we really do not know how long Adamah was alone before Eve was made. But the event of Eve's entering the scene would have been able to be done in a fairly pristine environment and likely included the use of earth-slime much like with the other creatures and Adamah.

From a Rib

Since God had Created Adamah in the Image of God, that is to say in the image of the **I Am**, and since the Creation text states that both Adamah and Eve were made in that image, we can safely assume that God wanted Adamah's companion to also be made in that image, but also in the very specific image of Adamah so as to keep the image highly consistent and pure to the **I Am Image**. The idea of having basic design data being infused into a slurry of nutrient-rich earth-slime or water for the other creatures, allows for a great amount of diversity. But diversity is something that the **I Am** Creator apparently did not want in the case of Creating Adamah as is implied in "in the image of God he Created him: male and female he Created them." One way we make duplicates in our modern era is to clone animals by taking a cell's DNA from an animal and infusing it into the embryo or egg of the intended clone. And it appears that is similar to what occurred with Eve. "Then the Lord God cast a deep sleep upon Adam: and when he was fast asleep, he took one of his ribs, and filled up flesh for it. And the Lord God built the rib which he took from Adam into a woman: and brought her to Adam."

At this point in staying with the slime-slurry concept, we would have the same with the exception of the use of lightning's electricity to pass the DNA sequence, because in the case of Eve Adamah's rib was used for the infusion of information from which Adamah's DNA would have been utilized to make Eve. This would produce the additional version of the creature type referred to as "man" which now adds "woman" to the Creation list.

Now, since Adamah was the first of the creature type called "man" to be Created, he was very specifically made and was alone. There were no plurals when Adamah was Created as there likely was with all the other creatures. It was one creature "man" in the image of the **I Am**. When Eve was cloned from the DNA from Adamah's rib, Eve was also of the creature type "man". Translation is a nuanced task and the original text was not written in modern

English. However, it is interesting to note that "woman" when dissecting the word based upon the Hebrew alephbet letter utterances, that the English "w" sound is the same as the Hebrew "vav" or "vaw" or "waw", which has the meaning of *attach*, or *together*, or *add*, or *also*, or *with*. Thus the term "woman" would have the meaning of "***also man***" in that context. Eve was likely Created through a DNA transfer process (a cloning process) and was therefore made like Adamah in the image of the **I Am** utilizing the DNA of Adam.

It is also important to note that since Adam was allegedly made from the "slime", it would indicate that it was wet and included dirt according to the Creation text. Now, this in itself is not a big deal, but if you consider that the Hebrew "mem", our letter "M" means water, and that the Hebrew letter "nun" or "noon" our letter "N" means life, or alive, or living, it makes the term "man" have a meaning of "living water", water which is coincidentally roughly sixty percent of our body weight. We are in fact, undeniably "living water". This is perfectly consistent with the Bible. To add to this, the term "hu-man" in the same line of thought in ancient Hebrew indicates "vessel" *with* or *of* "living water". So not only are our bodies technically vessels, but we are also scientifically living water. To add to even more this thought, our soul is embodied in our body vessel and that soul can be filled with the True Living Water of the spirit of God that the Christ offered to us, but I digress. Eve, who is "woman" is also "living water".

Interestingly, the terminology "kind" is not used in reference to "man". Man is simply "man" in the Creation account, and since "man" means "living water" and is in the image of the **I Am**, Adamah and Eve were unique amongst the creatures. From a scientific perspective no one should have any problem with the idea that Eve was basically cloned from Adamah's rib and thus is also "living water", which is to say wo-man or ***also***-man.

Woman is the crowning achievement of perfection of the **I Am** in beauty having been given the ability to Create life in the

Image of the **I Am**, which is all truly amazing! And that occurs even after the fact that she and Adamah fell short in the Garden regarding the Tree of the Knowledge of Good and Evil. You can read more about the details of the fall of man in the book *Understanding The Bible - The Bible How-To Manual* AND *The Things We Don't See* and in *Understanding The Church - Upon This Rock I Will Build My Church* and also in *The Science Of God Volume 5 - Boats, Floods, and Noah - The Deluge*.

Here is another ancient Hebrew point to consider in this; "he took one of his ribs", that is to say that God took one of Adamah's ribs and with it "the Lord God built the rib which he took from Adam into a woman". Rib indicates *represent self-existence*, in other words, something was taken from Adamah that would represent himself existing or would exist because of himself. Here is the text again for your evaluation: "Then the Lord God cast a deep sleep upon Adam: and when he was fast asleep, he took one of his ribs, and filled up flesh for it. And the Lord God built the rib which he took from Adam into a woman: and brought her to Adam."

Due to translation nuances, it is possible that it was not actually a physical rib, though it very possibly was, but whatever from wherever was taken was removed "and filled up flesh for it". And from that Eve was formed. Now, in a childlike manner, we can imagine that God physically sculpted the rib into a woman with God's hands, but since God is pure thought, that is highly unlikely. However, when we run our thoughts along the line of cloning Eve by using something from Adamah's body containing his DNA, then this phrase "the Lord God built the rib which he took from Adam into a woman" makes a whole lot of scientific sense.

How would God have done this process? It is very possible that it was also lightning. I Am throwing this out there only as a suggestion, but Adamah could have possibly been struck by lightning knocking him out as in "Then the Lord God cast a deep sleep upon Adam: and when he was fast asleep, he took one of his ribs, and filled up flesh for it" while the wounded area from which the "rib" was taken was healing shut, meaning that God "filled up flesh for it". No matter how we view this, Eve was made from Adam's rib and in our clumsy modern verbiage and thoughts; Eve was in some way "cloned" using Adam's "rib", likely through the function of lightning striking Adamah.

Construction Differences

We live in a world that, for whatever reason, trends towards corruption until total destruction occurs and we are then forced to rebuild society. This occurs due to our affinity to follow foolish trends. The differences that comprise men and women are many and they are obvious to any honest conscious observer.

Evolution theory would have us believe that somehow a single primitive amoeba morphed generation upon generation upon generation into a viable creature that could self-replicate without having to mate, which is similar to the way cells divide in order to multiply. But this in itself is quite a stretch of the imagination when considering the detailed intricacies of any single cell's inner workings that are all **required** for a single cell to self-replicate. In the evolution dogma, amino acids somehow went from amino acids to highly complex mechanisms all from no specific cause with no instruction. Ignoring the grand leap from amino acids to a fully functional cell, we can see that these newly formed cells could replicate themselves. Cells do this on their own quite reliably. But now we have an additional chasm to jump. Through what mechanism or influence did what we refer to as the "sexes" come about?

Evolutionists will present all sorts of organisms that might be very basic that resemble a lung or a heart, or, in this case, a penis or a vagina. Then from this they will in some way convolute their thoughts and somehow attach this through their morphology to creatures and then suddenly creatures need to breed in order to propagate. However, following the evolutionary flow, there is simply no need for this to ever occur. Using the evolutionary logic of the-survival-of-the-fittest, there is simply no cause that would create the sexes to ever have arisen since in the evolution theory they were not ever needed because cells already self-replicate and all descendancy would inherit that ability.

Male and female are different, which is clearly evident in each sex's design. There is no arguing this point. The first **I Am** replica man which was called "Adamah" or "Adam" and the second **I Am** replica was a clone of the first called "Eve", and they were designed to come together to Create more replicas of the **I Am**. Their physical gender parts are designed to work in harmony and one will not produce anything without the other. Here on our Earth, this has been proven billons of times over the years, it is one-hundred percent reliable that when the egg from the woman has information added to it from the DNA seed from the man that a new **I Am** image immediately begins its formation. It is true that some of these **I Am** Images are not able to come to maturity, however, they are no less images of the **I Am**.

Man and Wo**man** are complimentary Creations that, in the long-term, cannot survive without each other. The affinity of man to woman and woman to man is so deeply natural in us all, that the only thing that can thwart it is deliberate corruption by the evils of society. Science, nature, The Bible, logic, commonsense all agree vehemently on this issue.

Choice of Breeding

If you have ever read some of the original works of the evolution proponents, you will find that there are many phrases used by Darwin, such as "*there can be no doubt that...*" upon which all of the subsequent thoughts he had are built. And this is okay for an author to say, but what we have to realize is that the "*no doubt*" is the prerogative of the particular author, and if that author's "*no doubt*" is incorrect, then so are the theories that are built upon that "*no doubt*" assumption.

When speaking of evolution, the idea of natural selection is often implied and used to theorize transition from one form to another over many generations of that bloodline of creature. The theory is that the animals select mates with the most attractive traits, such as strength and beauty etc. and through this and many other environmental conditions they will slowly speciate into an entirely different species. This sounds logical, but it is not proven other than variations within the particular kind or species that we see when crossbreeding animals. This whole debate in large part comes down to the limits that define a species, and those limits are quite flexible in the evolution world. This will be denied by many, but you need to only briefly read the works of evolution and you will quickly witness inconsistency with regard to species definition scattered throughout the works of evolutionists. In one case we can take a creature and loosely define it as a species, but then when needed, we tighten the definition up for another creature to make things look legitimate, even if it is done in oblivious innocence. So if we need to connect two somewhat similar creatures that otherwise would not connect in that way, we simply broaden the scientific scope of a species' definition.

Regardless of whether they were Created, or evolved from the initial amoeba, it is truly amazing that creatures mate only with other creatures of their kind. With Adamah and Eve they did not have any other possible choices because they were the only ones made in the form that they were made. But when we consider the modest evolutionary deviations that can occur in man, we have to look at some of those that we consider features that we find less appealing and wonder why we feel that way. This appears to be something that is natural and, as far as we can tell from history, it has always been this way. Our natural desire for a healthy strong and attractive mate is something that is more determined by women than it is by men. But if a man finds a woman unattractive, he can then decline and thus stronger and more attractive instances of man and woman will be more common in the world. This is certainly true if you view the world in an honest manner, rather than in the way modern media at the time this book was written tends to view beauty and strength.

Chapter 7

Multiply by Dividing

Adamah and Eve were commanded to begin to propagate when God said: "increase and multiply, and fill the earth, and subdue it, and rule over the fishes..." etc. Multiplying by dividing are essential functions of life, without which we, that is to say mankind or more appropriately "man", would no longer exist. Before they could multiply, they had to come together and be joined to make one flesh. We can assume at this point in the Creation text that God was done Creating creatures and from this point forward the creatures would all have to self-propagate or their kind would have perished.

Now, if we take the text at its word from an authoritatively translated Bible, we must take the text for what it says, and with a pinch of common sense, we can evaluate the way nature works and compare that with the Creation text in order to get a logical and realistic picture of what likely occurred. Since the text did not mention "kind" when Creating man, we are going to make an assumption that is consistent with what we see in nature today. "Man" has very little deviation from the original design, and any male human can breed with any female human and produce offspring provided that the female is within her child-bearing years and the male is fertile. Then, when the woman Creates life, the new life in the image of the parents will not deviate much at all and the child will have attributes of the parents, yet will be a unique version of the **I Am Image** we call "man". On rare occasion offspring will be ill-formed, usually due to environmental forces, including parental habits. When this occurs, the ill-formed offspring will either die too early in life to produce offspring, or as is often the case, they will be unable to find a willing partner with whom to propagate. For these reasons man and woman have stayed reasonably consistent in form and function as far as all evidence has shown thus far.

Repetitious

When the female's egg cell is fertilized by the male's sperm, these two components are joined together to become one flesh. This occurs rather quickly once the sperm cell is able to penetrate the female's egg's outer surface. Almost immediately upon merging and becoming one, the newly formed unique cell containing all of the required instructions begins to divide, and through this division the cells quickly multiply to astonishing numbers for the given creature, in this case it is "man". This pattern of Creation is unparalleled in complexity and consistency regarding anything we as man and woman have done in all of our existence. Plant seeds share this same extreme consistency.

No matter what kind of plant or creature, the offspring is always after its kind. And this does not preclude the ideas of evolution because slight deviation can occur in each generation of offspring, and if that deviation is strong enough then the creature type will eventually pick up those traits as defining traits. We see this in animal husbandry and in horticulture where the given kind will be crossbred in effort to influence certain traits. Yet the newly divergent creature still remains within its broader "kind" as set forth in Genesis. We see no deviation from this model, and the pretend charts of evolution are invented imagery with tremendous speculation in order to bridge the gaps that are strewn about in the pictorial evolution charts. The same type of deviation found in animals occurs in man also, however, it is to a far less degree, which we can witness anywhere we look.

The vast accurate repetition of cells, along with those cells somehow understanding their specific task, is undeniably ingenious and is also quite astounding to behold no matter how it came about. If this occurred without premeditated intent as evolution proposes, then it is a fluke with odds so low that all of the atoms contained in all of the stars in the heavens could not represent the ratio for the extremely low odds of evolution to have occurred as proposed. There are all sorts of nonsensical terms thrown about in this debate such as "irreducible complexity" and "natural selection", and while those do have some validity, they fall short of understanding the magnitude of the task of bringing forth *any* life.

This idea of DNA being instructions is easy to understand, and we can just say that DNA tells the cells what to do. I can accept that just fine, but I want to know how this happens. How does a seemingly identical cell know that it is the cell of an eyeball verses being the cell of a fingernail? We may be able to understand this at some point in the future, but we will likely struggle to explain how it evolved to the point of cellular knowledge so as to have the cell carry out its specific task. From a mathematically scientific standpoint, the odds of cells knowing their place in the body is extremely low, yet they do so trillions upon trillions of times per second all around the world. To get from chemicals hit by lightning to an amoeba is in itself a longshot, but to get from an amoeba to something simple like a cell of a single hair follicle is on an entirely different level. Now, increasing the complexity of the target part to something like the ear drum takes things to an entirely new level, a level that is truly unable to be comprehended regarding the low odds of it ever occurring through **any** form of naturally selected evolution.

Evolution proponents typically will take a single bodily organ and then find examples in nature that appear to be of a similar design as mentioned in a previous section regarding gender, and then claim that this is confirmation that the particular organ could have come about from that type of creature. From an evolutionary perspective I can accept that, though it is a major leap of blind faith. We can take organs that are more complex that have many functions and then make the same overreaching arguments so that it seems plausible in an evolutionary model. However, it is when we compound these items that this idea becomes ridiculous. Now, evolutionists can make the claim that these things all developed in conjunction with each other, but that only serves to reduce the odds of it ever occurring at all. Allowing a bit of intentional framework to have been infused into the earth-slime is far more rational of a theory than is amoeba-to-man evolution.

Where science goes off the rails is in the dissociative reasoning. If you are able to disconnect two completely incompatible thoughts that would otherwise be required to be connected in order for the theory to prove out, then you can make any theory work in your mind. This problem of dissociative reasoning occurs due to lack of standards in definitions and from failure to use true and pure logical reasoning.

Definition of Terms

We have been taught to have a certain amount of respect for the science fields, and to a point rightfully so because science has brought about many wonderful advancements. However, the standards are vastly different depending upon which branch of science you study. It also depends upon whether you are only a theorist, versus being an engineer who must make things work properly so that people don't die while interacting with what the engineer designed. In astrophysics—probably the least accurate of all sciences—if the engineers that design and send rockets into space used the same rationale that the big bang theorists use there would be a whole lot of dead astronauts. Similarly, evolutionists can invent wildly out-of-touch claims that are typically extrapolated from microbiological morphing, which they claim to be evolution and I suppose it is technically evolution. However, when the microbiological matter is altered through sheer force, it still remains in the same biological realm it originally was in.

Any thought of altering DNA sequencing and then reusing the modified DNA cannot be considered evolution because it is done with a tremendous amount of intelligence via the lab technicians and is mutated by force and is no longer "natural".

The evolution-versus-Creation debate fell off the rails the moment someone decided to define "kind" beyond the parameters set forth in Genesis One.

Defining a "kind"

In discussing evolution, you will hear evolutionists demand a definition for the term "kind" as if mocking it in a way that makes it sound as if it is not a "scientific" term, however, as mentioned early in this book, the term "kind" was translated from the old Latin term "species" found Latin Bibles of antiquity. This means that, today, "science" is using old Biblical terms to define the evolutionary groups. The old saying *"a rose by any other name is still a rose"* applies here. It doesn't matter what we call something, the naming of the thing does not change the thing—it still is what it is. To make this more personal, if you decided to change your name, would you then suddenly become a different person? Obviously not, though some may pretend so.

Based upon the Creation text in Genesis One, we can make a very safe assumption that it is now and was always intended to be very broad, thus allowing for a great amount of deviation. The "kinds" listed are terms such as "birds of the air, fish of the sea, cattle, creeping things, and beasts". There is no more of a distinction than that, and anyone who tries to back a Creation proponent into a corner by forcing them to Biblically define thousands of species is ignorant of the actual words of the Bible's Creation text.

Hebrew, Latin, and Greek

Science's main job is the splitting of hairs in effort to make distinctions in things so as to better understand them so that we can utilize that information for the betterment of man and ultimately to know God better. This is why ancient languages are often used in medicine and science, because they do not change as do modern languages, thus their meaning tends to be far more stable. And this is why the term "species" has been selected as one of the defining terms to replace the English term "kind". For this reason, Hebrew and Latin and Greek languages are often used in medicine and science, with the Hebrew being least used of the three for such purposes. It might interest you to know that the term "kind" is rooted in the term "kin" like in the phrase *next of kin*. And "kin" is connected to the Latin term "gignere" meaning to beget. So basically, all of these words come back to the same thing where the creatures would propagate after their own "kind", meaning that the babies will grow up to be just like mom and dad. To sum this up and parallel the quote about a rose: a "beast" by any other name is still a "beast".

Chapter 8

Forming the Vessel

Most people are too busy living their lives to study these sorts of things, so books like this are written to consolidate the information and weed out the garbage in effort to deliver the concise details needed to easily understand the topic. What most people tend to not notice is that it is the livelihood for most of the people in the evolution industry who deliver much of the information to you about this topic, and they make their money by constant informational updates typically in the form of books and "documentaries". There is nothing wrong with doing that, this book is similar in that way, but when the interest in the subject drops off, then a new more-fantastical book will be written with ever increasingly outrageous claims. And again, we should have no problem with this, until they attempt to delegitimize other books such as the Bible, thus causing people to turn away from the **I Am**.

As the various theories and deviations of those theories evolve faster than the evolution that they profess, they are formed into a mindset that fills the heads of unsuspecting youth with what amounts to lies and theories that are spattered with things that are true to make the overall theories appear true throughout. You will notice this when you look for it where an evolution theory is cobbled together with chewing gum and masking tape. But then when you point out any of the many faults in the theory, they will come back with the only parts that are true which they have either dishonestly or incorrectly associated with a given aspect of their theory.

Evolution is obviously real and it obviously occurs, the only question is its *scope* or its parameters of allowable deviation. Evolution as presented in most "documentaries" and most ape-to-man type books is filled with inaccuracies having more holes in it than can be counted if you can keep count of even the rate of changes in the theories.

By Design

When discussing the evolution-versus-Creation topic, many evolution proponents get very upset when terms like "design" or "intelligent" are used in explaining Creation. Those terms are these sorts of open-ended vague tactics that Creation proponents use to broaden the scope a bit. They allow for a broad range of possibility regarding methods of Creation. Thus, some Creationists will say that "yes we evolved, but God put the mechanisms in place through intelligent design". This is a fair enough approach, especially when considering all of the far-reaching statements made by the evolution side of the debate, yet this is a cop-out regarding Creation.

In the last chapter, definitions were discussed regarding defining terms like "kind" and "species" used in the Bible. The old Latin term "species" was hijacked by evolutionists some hundreds of years ago, and along with it was the commandeering of the word "evolution", which means little more than "change". The Creation people dare not use the term *evolution* for their own use lest they have the evolutionists descend upon them with the force of "natural selection".

To not see the obvious design in the inner workings of all of biology is to be deliberately working with eyes and mind closed off to all Truth. In evolution, when similarity is noticed in general function, it is always attributed to a similar thing having come from the creature's evolved lineage. But while I suppose that possibility exists, the engineer in me cannot accept association in that way as having descended from something else for being the reason they are similar. Blood, bones, and flesh are just good inventions and they are incredibly versatile. If someone invented woven fabric to make a shirt, why would they not also use that same method of making fabric for use as curtains, or cushions, or pants, or the sail of a boat?

Design is so obvious in every single part of nature that one would have to be a fool to deny that deliberate intended design exists within every aspect of nature. As mentioned in the book *The Weightloss Repair Manual - Lose Weight While Sleeping*, "We are not just some random freak chance beings that randomly formed because some lightning bolt hit some amino acids and then formed us over billions of years. We are Created 'in the image of God.'" If you can accept that simple fact, then you will quickly begin to see the design in our bodies and everywhere you look and you will immediately begin to understand that there is specific purpose for every part of our bodies. For instance, even our ailments are obvious warning signs that are desperately trying to tell us that we need to change our habits and direction, but sadly we often fail to listen to those well-ordered indicators.

Versions of Slime

Adamah is said to have been made from slime and is also made of dirt, thus we can safely conclude that the "slime" was what we think of as mud or as it was referred to earlier in this book as, "earth-slime". With the plants and birds, and other animals it was the sea and the earth that "brought forth". Since there is great diversity in the very few but very broad kinds listed in Genesis, we can assume that this same diversity was not so with Adamah. He was a unique Creation, and while the other creatures likely where also borne out of some sort of slurry similar to the way Adamah was, it appears that there was only one Adamah. The earth would have had abundant and varying slime holes that would have varied due to the surrounding available nutrients. If you infuse basic information into that slime, the diversity of results would be as diverse as the conditions present around the world at any one point in time.

This same description was not used for Adamah, his Creation appears to have been a controlled single occurrence that was specifically tended to by God. We can safely speculate that this would have been done in effort to maintain a God-like form that would be passed down to all subsequent generations of "man", which is consistent with the real world that we witness at this very moment that you read this, as well as with the statements in the Genesis One and Two text.

Replication Errors

The ability for accurate replication in the microscopic realm is truly astounding. There is very little deviation, and the smaller you go the simpler it becomes. But on the larger scale with man, we do find replication anomalies in the form of birth defects which are often indirectly due to environmental interference. One such birth anomaly is "microcephalus" where the person's skull is considerably smaller than is typical. These things happen, and those who are the recipients of such anomalies often die at younger than normal ages. Such birth anomalies have been occurring for thousands of years and those offspring were at times altogether rejected by their parents and community.

There have been many geological digs around the world and countless skeletal remains have been unearthed. Most are attributed to typical modern human structure and ethnicity. However, on occasion an outlier is found, and when it is, it is **big** evolutionary news! What they typically won't tell you is that the stories behind these outliers are all utter fabrication, including the timeline and the full form and the way the newly found skeletal structure allegedly might have looked when it was alive. Are these stories possible? I suppose someone could argue that the stories are plausible, but are they likely? Not so much. These invented stories are built upon centuries of twisting facts. One day soon this entire house-of-cards will come tumbling down.

I truly have no problem with people making far-reaching claims about their archeological findings. What bothers me is that this is presented as if it is "settled fact". The "missing links" that have been found are all still missing. If you have not been studying this subject in any deeper manner, then you are just fed the garbage that comes to you through the documentaries of consensus. Most of these stories are more imaginative than Greek mythology. When an outlier is found, some of those outliers are likely due to birth anomalies rather than evolution. If you ever get the opportunity, do a bit of digging and find pictures of the various "transitional" remains found that supposedly link man to primates, and then also search for birth defects in modern-day humans. You might be surprised at what you find.

When removing all of the fantastical artists' renderings and the sculpted pieces used to fill in the many missing parts, the extremely fragmented remains wouldn't even fill a single casket. If a normal skeleton is found, it's no big deal, but if one is found that is out of the "normal" range due to possible birth defects, it is then an anomaly that will get a grand amount of attention and be dated to millions of years old as it is claimed to be a missing link even though it is likely nothing more than a birth defect.

Eukaryotes, Prokaryotes and More

Because terms like "species" are broad and can be used for many evolutionary levels, there are other defining terms such as prokaryote, eukaryotes, phylogenesis, and many more that target certain levels, of development and can describe or define differences in the creatures. All of which ultimately comes down to trying to define "species", or, as most modern English Bibles put it, "kinds". This splitting of hairs is not wrong or bad in any way; rather it is abused and misunderstood.

The part of this that is rather peculiar is that the evolution proponents who profess that evolution is an ongoing process that never ends and that there is no specific sudden form change, and that each "branch" of the evolutionary tree developed slowly, are always finding "new species". No big deal, right? New discoveries are always intriguing. But the issue here is that evolutionists insist that there are no distinct lines and all changes were smooth seamless evolutionary transitions, yet it is the evolutionists who insist that the Creation proponent define "kinds". A "kind" is a broad and loosely defined type of creature, such as a "beast", that has an incredibly broad scope, there is no more or less to it than that.

Chapter 9

Preserving Creation for the Long-Term

One of the longest and ongoing debates is the evolution-versus-Creation debate, and as mentioned in a previous section, that debate continues partly due to the misunderstanding of words and definitions. When you study the fossils and remains of the many creatures that have been found, what you find is largely what you see today in nature regarding animal forms, with the exception of obvious extinctions, such as dinosaurs. The layers of rock and dirt that have entombed the various forms have in many ways cleared up questions, but have also caused many new questions. If we found no deviation whatsoever from the skeletal structures in modern creatures, then full scope evolution itself would be entirely extinct. However, since we do find minor amount of deviation and also find many enormous sized extinct creatures, it does lend somewhat to the idea of full-scope evolution.

The fossils and other remains of creatures from eons past are a true gift for researchers or anyone else who has an inquiring mind. Without this history buried safely beneath our feet, we could invent any theory of how the creatures came to be, and no matter how absurd, no one would have any solid evidence to refute our absurd claims. Yet even with this vast historical geological record, we still tend to invent some far-reaching outlandish theories like ape-to-man evolution.

The Bones of the Matter

Researchers have found many human remains, and in general we discount most of them as modern man. However, on occasion we find peculiar human remains, and when we do, they are usually attributed as "primitive man". Such conclusions may appear true based upon pop-science's analysis, but we must also consider what we think of as the birth defects or abnormalities mentioned earlier. For instance, there are those born with non-normal skeletal structure, such as microcephalus where the cranium has not fully developed and the brain compartment is much lower profile and is smaller than as is typical in man. If such a skull was to be found and is claimed to be of ancient origin it could potentially appear as an "evolutionary link" in the mind of the finder of the remains.

All human remains found have to be interpreted by someone, which is typically done by the person who found the remains. If a researcher is looking for certain speculative features, then an outlier skeletal find that lends to those speculative features is the only find that will be acceptable to that researcher. Thus, if typical human remains are found they will be discounted as if they are more recent regardless of how deep they were buried, but outliers are immediately be accepted as potential evolutionary links even when found on the surface of the ground.

There is tremendous deception in the analysis of fossil and human remains finds. Many of the remains of human-like creatures are misrepresented either deliberately to further a theory, or ignorantly doing so, thus causing them to believe that they have actually found the object of their quest. The way we decide to interpret the evidence we find will determine what we claim that the evidence means. If we want to prove the theory of evolution then that is what we will claim the evidence is showing. But our claim does not make the proposed theories that are stated in full-scope evolution true.

Phylogeny

I once read a document regarding speciation with the phylogenic chart, and it was proposing a challenge to anyone brave enough to accept the challenge. The point of such a challenge is for anyone who is a Creationist to attempt to draw distinct lines of demarcation where each species or kind can be clearly established that would ultimately defy the theory of continuous evolutionary transitions. Taking such a challenge without accepting truth is a fool's errand. This is a trap for those who have not read the Creation account in an authoritative Bible and therefore believe wrongly about Creation and how the Creator likely accomplished it all.

I understand how people want to defend God and all, but God needs no defense. And "defending God" to push your own agenda is, quite honestly, selfish and foolish. An authoritative Bible is really quite clear in its statements; however, because we are so blinded by our religious presuppositions, Genesis is often misunderstood. As mentioned in previous sections, the "kinds" or "species" mentioned in the Bible are extremely broad and few, and they allow for many varying forms within each broad "kind". So, what occurs when people accept this ridiculous phylogenic challenge is that they feel compelled to try to draw distinctions where no distinctions are stated in Genesis. Here is the text:

"... God also said: Let the waters bring forth the creeping creature having life, and the fowl that may fly over the earth under the firmament of heaven. And God Created the great whales, and every living and moving creature, which the waters brought forth, according to their kinds, and every winged fowl according to its kind. And God saw that it was good. And he blessed them, saying: Increase and multiply, and fill the waters of the sea: and let the birds be multiplied upon the earth. And the evening and morning were the fifth day. And God said: Let the earth bring forth the living creature in its kind, cattle and creeping things, and beasts of the earth, according to their kinds. And it was so done..."

Let's break the "kind" listing out of the text:

Let the <u>waters</u> bring forth:
the creeping creature having life
the fowl that may fly over the earth
the great whales
and every living and moving creature
and every winged fowl according to its kind
and let the birds be multiplied upon the earth

Let the <u>earth</u> bring forth:
the living creature in its kind
cattle and creeping things
and beasts of the earth according to their kinds
and cattle,
and every thing that creeps on the earth after its kind

There are a couple of ways to interpret this "kinds" list. The first interpretation is to assume that the mention of "kinds" is intended as explicit detailed kinds like *robin* and *sparrow* and *pigeon* etc. The other way to interpret this is that each is "kind" is a broad-scoped group where a "kind" is a "beast" or "every living and moving creature", and "every winged fowl" etc.

The Bible's listed "kind" distinctions are very broad and few, and the statement "each after their own kind" makes no explicit affirmation as to any subgroups within a "kind", though there likely were variations included within the listed "kinds" in Genesis One. However, regardless of there being a broader "kind", or further refined groupings within the broader "kind" scope, the statement that they reproduced "according to their kinds" holds true in Genesis and in the evolutionary model, and of course in reality to this very day.

So if you are debating on behalf of the Creation model, do not take the bait by attempting to expound on the "kinds" list beyond those actually listed, such as "beast, winged foul" etc, and further, within the limited list of "kinds" actually stated in Genesis One, you need to go back to the Latin or Greek words and examine the deeper meaning of the words that were used to designate the only stated "kinds" that are actually mentioned in the text. The scope of the Bible's Creation text is extremely broad and very interesting when you remove the late post-Reformation rose-colored Christian glasses that have blinded so many people and trapped them in layers of phylogeny.

Long Age Layers

While the actual earth embraces and so eloquently encapsulates history for us in such a concise and logical manner, that particular history is typically greatly misunderstood or misrepresented by man. The layers we find, do in fact, tell us a great deal about the history of the creatures, but what period are they revealing to us? This is where scientific confusion and speculation runs rampant. Defining of "evolutionary ages" is similar to the defining of species, which is a very subjective task. Who gets to say how old a particular layer is? You might be surprised if you dig deep, and I mean dig really deep into the "geological column" to understand how the ages are determined.

While the layers' ages are largely speculative, what we do know using basic logic is that the layers nearer to the surface were deposited after the layers that they sit on top of. In some very rare situations it is possible that a lower layer may have been pushed horizontally on top of layers that would have been deposited at a later time, but those occurrences are very rare and typically very obvious.

What we do know is that the layers are relatively old and that they have entombed a vast array of creatures, most of which are waiting for the next group of researchers to unearth them.

Written Preservation

We are told to accept modern scientific writings and disregard the Bible's Creation account. I agree with this when it is one of the aforementioned late post-Reformation Bibles with horrible translation, but I take issue with that perspective when it is an authoritative Bible version. The Bible, much like the layers of rock that entomb so many creatures, is a record of history, but if that history has been tampered with then we have a problem, such as the problem that occurred with some of the late post-Reformation Bible versions just mentioned.

We must make a distinction between fact and speculation. It is a fact that there are layers, and within those layers people have been finding fossils for a very long time. It is also fact that nearly everything in the Bible is a historical account written down around the time it occurred, with the exception of prophecies and the Creation text. The Creation text is one of the most defiled texts out there and has been tampered with so many times it is difficult to keep track of. There are two books that are very helpful in analyzing the Biblical Creation data, *The Science Of God Volume 5 - Boats, Floods, and Noah - The Deluge* discussing the flood of Noah's time and *Understanding The Bible - The Bible How-To Manual AND The Things We Don't See* which explains Bible versions and which ones are best for more detailed research and why that is so.

The issue of Biblical translation sort of fades into a gray cloud of obscurity for most people, but to help better understand the necessity of using proper Bible translations, consider taking any respected book or written work of science and then having it reinterpreted by someone who did not properly understand the topic and instead misunderstood it in an entirely inaccurate light. There is no question whatsoever that as they attempted to re-explain the text as they translated it, the original respected scientific work would be greatly influenced by the translator's misunderstanding and would end by producing a scientific work that potentially leads people to wrong scientific conclusions. That is what occurred with late post-Reformation Bible versions; they have distorted the Creation text which has resulted in many people turning away from God due to telling people that the Bible is the "inerrant word of God" and then giving them a perverted translation claiming it to be "inerrant".

Any written history or written preservation is no longer preserved when it is incorrectly rewritten due to incorrect understanding; this is true even if the intent was innocent and pure. Wrong is wrong, and nothing will ever change that basic truth. To add to this problem, the same is true of the history that the Earth has written between the layers and preserved as a witness to the actual true history of Earth. The history of Earth has been perverted in much the way many late post-Reformation Bibles have perverted the Genesis One and Genesis Two Creation account. When scientists misunderstand the data encapsulated in the rock that they found and then expound on it in their books, they can only explain it the way that they understand it. Even if their understanding is entirely innocently incorrect, wrong is wrong and there are no exceptions to that simple truth.

Chapter 10

What We Are Committed To

Most people are more concerned with proving their point than they are with revealing truth. We commit to our errors with an astonishing level of voracity, unyielding in our quest to show others just how "right" *we* are. Our conviction and blind-faithed beliefs on matters such as a Creator versus evolution provoke much passion within us. And it is that passion that far too many onlookers respond to by interpreting someone's passion to be truth, but often it is not anywhere near truth. Typically, it's not that we are lying to them; rather it is that often we are simply incorrect in our own assessment of the data that we are then sharing.

Don't Be Fooled

Passions on the evolution side of this debate run so deep that many very detailed "documentaries" are made and books are written in effort to convince the unsuspecting public that the evolutionary theories are "settled fact". This even goes so far as to have artists do renderings of what the creatures that previously inhabited the bones allegedly looked like. But don't be fooled, often these artist-renderings are in the form of sculptures where they will take modeling-clay layer upon layer to make representations of the imagined muscle tissue applying it to the bones until they finalize the skin layer; then they proceed to add hair and eyes etc. Often these sculptures are *very* convincing, along with many additional artistic works that are done in the form of graphic representations with foliage and other evolved animals surrounding the central figure of focus. These depictions fascinate both young and old alike with their colorful, well-done, and imaginative flair. There are so many books showing such images surrounding the evolution topic that it is difficult to keep count.

But the evolutionists are not alone in such representations. Creationists also make similar final figures. Creationists have little need to study the bone structure because they are not making far-fetched claims. However, most of the depictions found in the Creationists' books and displays of animals are very similar to what the evolution crowd has concocted. It seems fair to say that the specific attributes of the creatures that the Creationists show are likely copying what the evolution crowd has already made, such as with dinosaurs. Are these depictions correct? Are they accurate? We may never know for sure, but what we do know is that every one of them is **_speculative_**. I Am not aware of any fossil finds that had skin intact well enough for us to know without question the exact skin texture of the creatures of eons past, especially dinosaurs.

The only thing we can do with our speculation is look at currently existing live lifeforms and then imagine what similar ancient fossils might have looked like while those creatures were alive and roaming the Earth. Maybe we have this all figured out and are accurate, but more than likely, we do not. Manipulative influence is buried in every bit of these sometimes-deceptive artistic renderings, which is especially true with the primate-to-man illustrations. If you happen to be interested in this subject and watch a lot of documentaries, you will have noticed typical looking skulls used as a base for clay modeling and you will find that two very similar looking skulls can have two very different outcomes. Often, through blindness or through deliberate deception, one sculpted figure will look like a typical human and another will have ape-like or primate-like attributes usually on the brow and around the mouth, plus ape-like hair placement. These compelling but unlikely compositions are photographed and subsequently printed in textbooks to perpetuate the accompanying evolutionary theories. Whether it is deliberately deceptive or ignorantly deceptive is of no matter because there it is for young children to see. Look for this practice and you will find it nearly everywhere this topic touches.

Debatable Points

There is no shortage of errors regarding the interpretation of data from both sides of the debate. On the evolution side there are no rules because that side is typically explicitly atheistic and has no concern about accuracy regarding the Bible or actual truth, or morality for that matter. The evolution side is only concerned with proving the theory of evolution. This might sound as if it is an attack on evolution but it is not, rather, this is simply what occurs.

The Creation side is no different in that they too seek to prove their "Bible-based" Creation theories. But on the Creation side the vast amount of errors is far more disturbing. Truth is of no matter on the evolution side because there is no core index for what "truth" is, so they will often divert to using the term "facts" and insist that their entire theory is based only upon "facts". But the Creation side's errors are far more troubling. It makes sense that they would want to stand by God's Creation and defend it, but as said earlier, God needs no defense. When the Creation side, who believe deeply in God, is in error because they have been preaching from the position taught to them due to the horrendously translated late post-Reformation Bible versions, it is especially heinous. The people who are supposed to be teaching about God and ***Truth*** are responsible for Creating more atheists than evolution is, and this problem is largely due to their own misunderstanding of the Creation text.

Many of the evolution apologists rightly criticize those who promote Creation because the Creation side is often at odds with reality *and* with the Bible. While the evolution apologists are typically somewhat more scientifically accurate and logical than the Creation side is, the evolution apologists are still incorrect about the greater part of their theories.

It is quite logical when you stop to consider how things came to be from nothing, where the whole idea of a self-Created Creator that became pure thought over eons of time while adding core concepts to the initial basic **I Am** awareness. In other words, it is obvious that a Creator exists. So on the issue of the existence of a Creator, the Creation side actually has that part right. It is their childish view of the way Creation was done and the time-frame in which they claim it occurred that are problematic where they sometimes invent really quite ignorant explanations for the geological record.

A Creator is not going to be arbitrarily Creating fully formed creatures with a magic POOF-like wand. A true Creator will be discerning and will use logic and what we think of as "common sense" at every turn of Creation. There will be no hocus-pocus big bangs spontaneously coming out of nothingness, and there will be no planets suddenly appearing out of thin air, and no magical Creation of "beasts". Nor will there be a man made of clay formed by God's hands playing in the mud. Creation is logical and, quite honestly, obvious when viewed according to well-translated text in authoritative Bible versions. The nuances of interpretation run deep in both the analysis of the Biblical text and of the geological record. The book *Understanding The Bible - The Bible How-To Manual AND The Things We Don't See* explains many of the Biblical interpretation nuances.

Animals Versus Humans

When considering the arrival of man, the evolution side has it all laid out and all creatures came from the lightning striking the liquid chemistry and created amino acids ultimately forming the beginnings of DNA. But here we already have a problem, because since DNA is the instructions for any creature, the DNA needs the creature in order for the DNA itself to exist. This creates a sort of circular dependency which is where any serious contention should come down to between the two sides of the debate.

Where the information came from that created the creatures is truly the core question in the debate, but that particular point is obscured by all of the petty bickering between sides. Is evolution correct and it all occurred over vast amounts of time, or is it more likely that The **I Am** God infused instructions of the basic "kinds" mentioned in Genesis, and then from that the "Earth" could "bring forth" varying forms within each "kind"?

Which is it?

Evolution has some very compelling points that most Creation supporters are unable to thwart due to their reckless adherence to bad Bible translations. Creationists are a bit closer with their "God did it" perspective, but are typically so very far off when it comes to the diversity that evolution points out. Evolution on the other hand struggles when trying to explain the superiority of man over animals. Man has *vastly* obvious advantages over animals, and since evolution claims that primates are the closest thing to man in the evolutionary tree, you would expect that apes and monkeys would dominate the animal kingdom, but they do not.

There are many animals that primates fear and will run from. The same can be said of man; however, man has dominion over the animals to a point where we are able to keep them in submission. That is not true of primates; they do not make all other animals subject to themselves. In a fight for survival, man in general will win over the animals every time. We will win because we are Created with a God-like nature and thus are able to premeditate and imagine ways to defeat any animals that might stand in our way, even if those animals are more powerful and/or more vicious than we are. We use our superior God-like intellect to proceed to build that which we imagined, and then use the tool we built to dominate animals that are more powerful than we are.

Regardless of which side of this debate you come down on, if you are unwilling to embrace and understand the opposition's arguments, you then will likely remain trapped in your own error. No self-respecting honest researcher or theorist is going to ignore the opposition's data and analysis thereof. Both sides have some strong arguing points, although the predominant Creation side's theories have been wanting in that area.

Chapter 11

A Seed of an Idea

Seeds make the world go round. Without seeds *nothing* could propagate "after its kind" as is explicitly stated so eloquently in the Bible's Creation text. No one on either side of the debate can argue that very simple fact. All seed material has the instruction-set for the particular "kind" variation. This undeniable fact is clearly stated in the Bible and is true in the evolution theory as well; however, in evolution it has no boundaries of a Biblically stated "kind". Evolution states that anything can morph into what would over millions of years end up being a completely different "kind", and in there is the core-overreach of evolution.

But seeds are more than physical constructs that propagate all living things. Seeds are also laid out in ideas. The basic premise here is that all seeds, whether they are physical seeds that will propagate life, or ideas seeds that will propagate thought, all contain vast amounts of *information*. And all information comes from thought. If you receive idea-seeds from someone, those idea-seeds will typically grow in your mind when you water them with ponderance. Hundreds of years ago someone pondered the similarities in various animals and considered how creatures in one area can be noticeably different than creatures in another area, yet they exhibit many similarities in structure and behavior and are considered to be of the same family group. Unfortunately, these particular information seeds happen to have evolved into the noxious weeds we today call "evolution".

Off the Rails

Thought seeds are where both sides can really come off the rails. Defining terms and making distinctions is where all problems in the evolution-versus-Creation debate arise from. When the seeds of evolution where planted, it stoked the engine of the evolution train to a point of it becoming an unstoppable runaway train. The evolution train plays on the vulnerability of the unsuspecting masses by redefining terms and inventing new terms that become defining labels causing people to think something is a thing that is not really a thing. You could think of this like geographical borders. Some borders are well defined such as major rivers and coastlines. Then there are less specifically defined borders such as mountains and massive forests or swampland, and finally there are man-invented borders that can be easily changed such as has occurred over the centuries with what is now called "Germany", for instance. The "scientific" evolution side of this debate has many such movable borders established in their vernacular terminology. The Creation crowd is considerably less complex and their fundamentals are also much older and more stable.

But both sides frequently leave the rails of truth to further each their own agenda. Establishing words such as *clade, phylogeny*, or more to the point of this book, *Ardipithecus, Australopithecus, Dryopithecus, Neanderthalensis etc* has an air of authority because many of these terms use ancient Greek phonetics, giving us the feel of scientific authority. Similarly, we also see this sort of thing in Creation supporters within the more scientific thinking groups. However, on the Creation side of the debate there tends to be more Young Earth Creationist who are outspoken than there are outspoken Old Earth Creationists. Young Earth Creationists typically are not as studied in science and the finer details of physics as are the Old Earth Creationists. Old Earth Creationists are usually scientists or professors, or people of logical minds rather than of passionate minds. Young Earth Creationists are often referred to as YECs and they believe that God Created everything in six twenty-four-hour days and that the Earth and all of the rest of Creation are only about six-thousand years old. Most of the Old Earth Creationists tend to believe that a big bang occurred and that we evolved, but that it was done through the command of God. The Old Earth Creationists are not in opposition to the atheist big bang evolutionists, except that Old Earth Creationists believe that there was intelligent interaction to get those things started.

The late post-Reformation Young Earth Creationists are very problematic scientifically, and since their stance is so absurd, it makes the Old Earth Creationists and the evolutionist appear correct, even though on many aspects most of them are not correct. But from a purely atheistic evolutionary side, both the Young Earth Creationists and the Old Earth Creationists are still grouped into a "Bible believing fairytale" status.

Lack of Definition by Too Much Definition

Science, by its very definition, is an occupation of *defining*. "Science" means to *know* or to *split*, and the idea of splitting is what a definition is. It is natural in us from our Creation onward that we want to know things. In fact, our desire to "know" is how "Original Sin" came about. Eve, and then Adam, wanted to be able to discern between good and evil, that is to say to have the mental ability to make the split between good and evil to know the difference, otherwise it is without distinction to them. We were Created in the image of the **I Am** whose very first act of thought was to Create new thought categories that define particular concepts. Each concept is a defining logical thought-border that allows subsequent thought to form. Our desire to know or split in order to define things is natural in us due to our being "made in the image of God".

To take definition a step further, consider the topic of *The Science Of God Volume 1 - The First Four Days* discussing the first few events of Creation. Definition was the fundamental core to all of Creation. Without definition, nothing could exist. God "divided the light from the darkness." Definition is an effort to organize and group things or thoughts into logical categories. This splitting, or "science", that the Creator did during the Creation events continued through the bringing forth of the creatures, all the way through to the making of "man". It even continued further to where "woman" was Created, thus making the "male" and "female"–two very distinct aspects of "man". And it was indeed a very notable and beautiful split or definition.

But are there limits to Creating distinctions? At this point we have to look at the distinctions made in modern science regarding evolution. How many erroneous gradations will we allow to be inserted into the actual existing found data? *Definition of species* has been the problem since the evolution-versus-Creation debate began. In fact, it is largely what caused the debate to even begin.

Going back to the "Phylogeny Challenge" mentioned earlier in this book, evolutionists attempt to force Creationists to explain every single variation that currently exists and also those variations that existed in the past. This is because the evolutionists have defined the various species in the way *they* see fit. But since defining species is more of an art-form with full license for Creativity, there are no true limits as to where species demarcation can be made, thus, too much definition Creates a lack of definition–It becomes a jumble of speculative nonsense.

Does the bird have a long beak, or a short beak? And are birds with long beaks different "species" than similar birds with very short beaks? Creationists often take this bait and will rightfully claim that those are mere adaptions of the "species" and that a wren is a "species". Yet even doing that is adding definition where the Bible states no definition, thus causing the Creation text to become convoluted in the minds of the onlookers of this debate. Unfortunately, most people never read the entire Bible. In fact, relatively few people have actually read Genesis One, and so they tend to rely upon those who speak about this for information and analysis of the topic.

With regard to the Creation-versus-evolution debate, evolutionists often demand that these species details be addressed, but according to the Bible's text, there is no need, and doing so defies the Creation text. There are fundamental divisions such as when God "divided the light from the darkness" and listed the base "kinds" in Genesis, but going beyond that is grasping at straws that do not exist in the text. In the evolution-versus-Creation debate there is no place for excessive definition beyond what is actually stated or actually found. Sure we can speculate, but we must not insist on what is not actually written or clearly logically implied in the text.

Thus Sayeth the Lord

There are many defenders of the Bible who will jump around in the text of the Bible in effort to bolster their debate points, quoting "Thus sayeth the Lord this", and "thus sayeth the Lord that", but to do so is often in violation of the Genesis text. It's okay to do this when it confirms the actual statements made in the Creation text, but using it to insert incorrect doctrine into the text violates the Creation text. The Creation text is unique within the entire Bible, as well as within all other ancient writings, as it is a brief but effective account of things prior to the making of "man". Since no man yet existed before Adamah was Created, there is no reference to man and nothing could have been affected by man with regard to the events listed in Genesis One that came before "man" was ever Created.

The Genesis One text is simple and straightforward, and it is only our own presuppositions of the meaning of the text that obscures our own understanding of its actual meaning. When discussing the Genesis text, we often assume the people speaking about it are the supporters of it, that is to say Creation supporters. However in the evolution-versus-Creation debate, it's often the evolutionists that are paraphrasing Genesis One. It is completely understandable that most evolution supporters defy their own interpretation of the Genesis One text because most of them were taught either directly or indirectly by some of the late post-Reformation Bible versions, and so they rightfully defy the nonsensical Creation accounts that are printed in those Bibles. This is perhaps the single most important issue in all of religion during our modern era. The book *Understanding The Bible - The Bible How-To Manual AND The Things We Don't See* is very helpful in this regard. Thus sayeth the Lord "my people perish for lack of knowledge" because they have been lied to by inept translators. Translation matters and Genesis One matters, and at minimum, it deserves accurate translation so that we all, at the very least, can read the honest account of it.

Chapter 12

Guided by the Layers

All around the world there is a great deal of evidence regarding evolution and the limits thereof trapped within the many layers of sediment. We find this evidence in bits and pieces as man traverses the Earth and excavates and builds. As we find these historical treasures, we strive to unearth them to reveal the facts. Then we take those facts and analyze them and present our own analysis as if that analysis is the actual found artifacts, which it is not.

It is a fact that we found something and that its structure matches that of a bird or a man or some other creature, and that's about all we really know about the actual artifacts. Any statements beyond that are pure speculation. The sedimentary layers reveal a great deal about Earth's history, but those layers have been contaminated in our minds by our own inaccurate reading of the actual found data. There is more information about the layers in *The Science Of God Volume 5 - Boats, Floods, and Noah - The Deluge*.

Rants Are Questions

If you closely follow the evolution-versus-Creation controversy, you will hear many rants from both sides. These rants tend to lean more to the evolution side because the evolution supporters tend to exhibit frustration and anger towards the typical inept Biblical perspective. And they have every right to do so when considering the young-earth perspective derived from the late post-Reformation Bibles. Since the Creation text has been so terribly perverted by far too many improper translation versions, people have a reason to get angry because they are ultimately being lied to whenever it is discussed. The inaccurate translations then cause people to turn away from both the Bible and from God. Those people then tend to turn to a religion that has more facts that actually line up with their logic than do certain convoluted Bible versions, and thus, they join the-church-of-science, who are no less religious than the Young Earth Creation supporters are.

But regardless of which side you are on, when the opposition is whipped into a frenzy and embarks on a tirade of a rant, what is really occurring is that they have unresolved questions that you are not answering for them. If someone insists that "God did it", or that "it all evolved" then you should be able to back that up with something other than saying that "the evidence is found in any eighth-grade textbook" or "God said so, the evidence is in the Bible."

We must be able to intelligibly answer their ranting questions that they often do not realize that they are asking within their rant. This same problem is true in most arguments regardless of what the argument is about. Rants or anger are born out of frustration from not being able to get satisfaction from the other party in the argument. Sure, sometimes people are just having a bad day, but generally we're frustrated because we are not getting answers to our underlying questions. Many times, this is more the fault of the ranter for being unable to properly form the question in their own mind in order to be able to properly ask it of their opponent.

But in the *overall* Creation debate, this is more of an issue with regard to the Bible supporters' ignorance of the topic due to the poorly translated Bibles mentioned throughout this book. With the evolution topic it is a bit different because evolution theorists have so long ago invented many a false premise that have now filtered into the hearts and minds of many people over the past couple of centuries. This influence has blocked the basic common-sense logic of the person asking the Creation supporters questions. It can be uncomfortable watching these debates because you end up listening to two opposing parties which both have incorrect analysis while both rightfully arguing mostly about the particular points that are incorrect within their opponent's theories. Much of this can be attributed to excessive definition where none is warranted and also lack of definition where it is explicitly required. This occurs because we fail to listen and generally won't even attempt to understand each other or the data sets given to us in the fossil record and in the Bible.

To add to this problem, not only do we misunderstand the actual facts of our own position, but we also misunderstand the opposition's position in general.

Our Found Species

In an earlier chapter, some of the various named "humanoid" primates *Ardipithecus, Australopithecus, Dryopithecus, Neanderthalensis* were mentioned. If these primates are accurately depicted, then they tend to fall into two very distinct broader groups, namely "beast" and "man". It is important when viewing such imagery to note that **all** of these pictorials are nothing more than various artists' visions of what the creature that inhabited the bones might have looked like. Adding expression to any creature image gives that image humanlike characteristics. We can do this on paper with nothing more than a few dots and lines and make simple arrangements appear happy or sad or angry.

Like this :)

or this   <: (

Adding these emotions makes what would be considered nothing more than an ape or monkey to seem to have human-like qualities. And then when it comes to the *Neanderthalensis* the opposite is true where the very human-like figure will be dumbed down to appear more animal-like due to suppressed expression.

It's peculiar that on the evolutionary side of this discussion you generally won't hear different "species" of humans mentioned. This is true even as it concerns the non-human alleged ancestors *Ardipithecus, Australopithecus, Dryopithecus*; instead they are all typically referred to by those and other specific scientific designations. It is the varying expressions given to the depictions of each creatures' visual rendition that blurs the lines that would otherwise have more readily allowed us to clearly mentally categorize them as either *primate* or man. Any truly honest researcher who examines the few existing fragments of these supposed ancestors that allegedly link to modern man, and does so in total honesty, will take issue with these very influential and very dishonest graphic representations. We should have no problem with these works of "scientific art", but we should take issue with the blurring of lines in this regard and with the insistence that there is a linear "factual" correlation between them.

To this day, there have been no substantial archeological finds that show any transitional man-like form that does not still exist alive to this very day. And further, some of the peculiar finds can easily be attributed, as mentioned earlier, to birth anomalies since those finds are rare—yet any normal remains found are not rare.

Race, Creed, and Color

Our international vernacular and our scientific vernacular are at odds with each other. When we speak of the animals, we have these very narrow limits for a "species" with various species of beetles or wrens etc. But with humans we don't use the term "species" or "kind", and scientifically that aspect is generally ignored by utilizing the Greek labels such as *Australopithecus* etc, instead we say "race". Science is at odds with itself in this as well. Evolution professes species and ever finer divisions at every turn, but with humans we don't declare Asians or Africans or Caucasians to be different species, yet the differences are clear and are equal to or greater than many of the differences that are between some scientifically defined species of a given branch of animals in the evolutionary tree, such a wrens. But this of course is determined specifically by whoever is detailing it.

Man is clearly grouped in the hearts and minds of everyone as being unique. With animals we have broad groups like birds, cats, cattle, etc. But we don't see the same level of variation in man. There are large birds like eagles and small birds like hummingbirds and yet we still identify both as "birds". There are large monkeys such as chimpanzees and smaller monkeys like the pygmy marmoset, yet we still identify them as "monkeys". These kinds of creatures all have similar intellect and general behaviors as their counterparts. But with humans there is no counterpart anywhere near our form or intellect level, and all of man shares that overall attribute.

The more liberal approach to science used by the evolutionists has trapped itself in a quandary of racism. To counteract their racist stance they give credit to the people of Africa as being the "cradle of civilization", but in so saying they are essentially indicating that the "Africans are not as far evolved" as they themselves are. Their evolutionary racist ideology is hidden by false credit given as the "cradle of civilization" and by a lot of shouting and belligerence whenever the race issue is even vaguely alluded to. In other words, you are not allowed to notice that many evolutionists are indirectly racists.

Man all around the world is distinctly different than any other creature. We define ourselves in terms of race, creed and color and in so doing, we create unjust "racial" tensions that should not exist. When we divide ourselves by our features it is not a bad thing until we somehow believe that someone with different features than ourself is somehow less than or more than we ourselves are. As also detailed in *Understanding The Church - Upon This Rock I Will Build My Church* as well as in this book, we are **all** made in the Image of the **I Am** Creator.

Chapter 13

Science is Fluid

Science changes with our understanding of reality, but reality itself never changes. Our circumstances might change and we might change our views, but facts like our Earth being round will never change, and the Bible says what it originally said and no poor translation will change the ***original*** words of the Genesis One Creation account.

We like to imagine that science has everything figured out, but it doesn't. In fact, science must change with every new revelation of truth or science will end utter defeat. The fact that for so long we have been told that "The Bible is just a bunch of invented tales" leads us to believe that nothing in it is true, and then when we read those poorly translated versions of the Bible mentioned throughout this book, it allows us to believe, first, that the translations are okay being presented as they are, and second, that none of it matters anyway, thus causing us to not even bother asking the tough questions about the Bible that we really should be asking. The information in the Bible, *especially* the Creation text, should be treated as a treasured artifact that should be studied and proven, but doing so requires more than just basic scientific contemplation of the content, it also includes forensic study of the documents themselves along the interpretation of those documents, rather than foolishly just believing that the entire particular Bible on your bookshelf is the "Inerrant Word of God"

Science changes and is supposed to work to better understand the unchangeable, which we have only just begun to scratch the surface of doing. The biggest mistake of evolution theory is that pesky assumption that evolution as proposed is "settled fact".

Success of Moving Theories

There are multiple theories on both sides of the evolution-versus-Creation debate. But the theories on the Creation side tend to be more stable in that whatever the theory is it doesn't change much, but that doesn't make it correct. Evolution does change, sometimes even during a debate. Moving theories like evolution are hard to attack or pin down because they keep changing things to avoid or hide the faults in the theory. This successful tactic works well in public forums because the slight-of-hand, or in this case the slight-of-tongue, is so subtle that most people in the audience will never catch the subtle contradictions and diversion tactics. The interesting thing is that many of the proponents of evolution probably don't even notice that they themselves are doing this. You can find diversionary tactics on the Creation side as well, but since the fundamentals don't change because they are written in the Creation supporter's Bible, there is little need to move the target.

A part of the reason that the evolutionary target moves is because it is speculative, and when someone points out a glaring contradiction then adjustments will quickly be invented to compensate for those gaps. The Bible indicates, in other parts of the Bible, that man is "Living Water", and in ancient Hebrew that is in fact what the symbols that make the word "man" indicate. And that is, in fact, what we are, we are mostly water and we are alive. This is consistent with the slime mentioned in the Bible as well. So, in that respect you cannot discount the Bible's statements about man in Genesis One or in the later parts of the Bible regarding this particular part of this topic.

The Evolution Task List

Evolution theory has an unspoken agenda or pattern of action that occurs at nearly every new discovery or observation. First an observation will be made and then immediate analysis will be done allowing speculative predictions to be made. Those are all fair steps in analyzing data, but from that point on the evolution analysis gets a bit, oh shall we say, creative? After predictions are made then retro-predictions are made. At that point, based upon those entirely made up retro-predictions, extrapolations of change occurring over very long periods of time are inserted. Then the potential effects of the imagined events in the extrapolated period are established and then assumptions are made of how the creature came to the form in which it was discovered as. Now, it is fair to say that there is little other choice than to speculate on these things because the past is past and we have to make some educated guesses.

But, that is exactly the point, evolution is all "educated guesses". However, in regard to evolution during debate, when mentioning "educated guesses", "biological evolution" will then enter the debate. Micro-evolutionary laboratory manipulation of cells causes the cells to change or evolve a bit to compensate for the forced manipulation from human interaction. The problem with long-age extrapolation is that because it is all invented and supposedly occurred in ages past, there is little opportunity to actually study it like we do with our microbiological lab experiments. This is especially true if the extrapolation is completely in error and could never actually exist. This is because it is very difficult to study something that does not exist.

Pay attention to these evolution theory actions that occur for nearly every artifact found:

- Observation
- Immediate analysis
- Prediction
- Retro-prediction
- Extrapolation over long ages
- Extrapolation of effects
- Assumptions/Conclusions

Pop-Science

As you hear the various evolutionary theories that are scattered abroad, be aware of pop-science. Pop-science tends to be far more belief-based than is real science. Pop-science is comprised of a small number of talking heads that are very skilled at garnering the attention of the media. They are provocative by nature and know how to market themselves. They are typically very snarky and will use ad-hominem attacks to cut down any dissenting voices. They are usually somewhat brash and condescending, but often actually likeable characters, which is typically why they are so successful. But, as always, none of that will make the many errors in their theories correct.

Real Science

Real Science is considerably more discerning than is pop-science. Real science deals with real life matters, such as sending rockets into space or creating new medical techniques that if done wrong will result in many deaths. If real science behaved like pop-science, we all would likely be dead already. Real science does not have the luxury of vast error the same way that pop-science's evolution or the big bang theory have. Real science puts fresh food on your kitchen table; it makes your refrigerator keep your food cold, it makes your vehicle's engine run. Real science doesn't make ridiculous claims by extrapolation back billions of years to unknown times.

Real Science is Godly, as in the name of this book set "*The Science of God*". Real science deals in facts and actual logical concepts. Real science mimics God and is from God. Where Pop-science, on the other hand, is borne of selfish-man and typically offers mostly lies and deceit for profit. And sadly, this pop-science trend hit the medical community in the early part of the twenty-first century resulting in many unneeded deaths and much regretted mutilation of youth.

Everyone must remember this as stated in *The Science Of God Volume 1 - The First Four Days*. "The Universe does not abide by our rules or laws, we abide by the Truth and the Order of the Universe and its forces, and there appears to be no avoiding this simple truth."

Man has a unique ability to ponder things and then logic through those things using our superior reasoning and intellect that has no logical explanation other than that we are "Created in the image of". When we depart from this then we become very illogical. It is up to each one of us to use our gifts of logic and reason to try to better understand Creation and the Creator. If the Creator does truly exist, then obviously the Creator is the foremost logical scientist that ever was, is now, and ever will be.

Chapter 14

What is "Life"

"Life", what is it? We generally look at life as it is some sort of living matter where it is able to grow and self-replicate to varying degrees. Plants are alive and animals are alive and people are alive. But life itself is more than just that. We say things like "I just want to live my life", but we already are living with each breath that we take and with every beat of our hearts. "Life" is obviously more than the clinical or scientific description of organic material. This is where science struggles between the *how* and the *why* aspects of life. Science really does not deal with the *why* because the *why* aspects of life involve intent. The *how* on the other hand is a bit more tangible because we can study *how* something works and understand action and reaction, but we fail to be able to scientifically answer the *why* it works part.

If you recall earlier in this book where some of the ancient Hebrew letters where mentioned, one of them was the *vav*. The *vav* or *vau* can be both our English letter "F" or our "V", and in addition to that, the Hebrew *lamed* is the same as our English letter "L". Due to a lot of migratory language phonetic translations, often the vowels get misinterpreted or left out, so it is best not to place too much emphasis on them. But if we take the *lamed* and the *vav* we can derive both the word *life* and *live* from those ancient letters, but also *love*. That is to say **lv** and **lf**. When speaking of the **I Am** Creator and of Creation, this begins to make a lot of sense. But do not be deceived in this. This is not some uniquely planned manufacturing of words, rather it is some fundamental concepts that man has assigned varying meanings to. They are in fact a single concept with multiple dimensions or intentions. *Love* and *life* are one in the same. If an **I Am** Creator actually exists, then the formative concepts that make up life are what make us alive and are one of the passions of the **I Am** Creator. From a Creation perspective, we are loved, we are life, we are alive. That is the *why*, but since science has not yet come to realize the entity of **I Am**, science has no means by which to understand *love*.

Gene Sequencing

Late in the twentieth century, we were able to begin to look at the strands of DNA that are verifiably included in living cells, and through that analysis we were able to "sequence" an entire DNA gene. As the sequencing techniques became better understood and computing became faster, reading the sequence was able to be done at ever increasing speeds. So, we were then able to read the entire sequence of gene nucleotides in very short order. But while this was an incredible feat to accomplish, we still to this day do not understand most of the sequence. Early in the twenty first century, science became even more proficient at reading the DNA but still struggled to fully understand it. A good comparison would be if you read a Latin language Bible, you would recognize many words and could reasonably accurately pronounce most others, but you would not understand the content or the context. It is possible that this complex DNA programming will someday be completely understood by us, but there is still a very long way to go, and when we do finally understand it all, what sort of horror will we do with that understanding? Understanding without wisdom is very dangerous!

The information in a single strand of DNA is based on the program infused into the earth-slime that Adamah/Adam grew out of, and our bodies' cells are extremely proficient at reading and replicating that data. This is true of all living things, but regardless, it is truly amazing especially when it comes to man. What took us years to figure out how to examine and record is naturally done in seconds by the trillions every moment of our lives, and done repeatedly to perfection. A single tiny microscopic cell in any person's body is infinitely more proficient at reading and understanding DNA and then applying that information, than is the entirety of the scientific community having access to the entirety of human knowledge. Cells and their DNA are actually a very amazing system that is unlikely to have ever self-created in any amount of time that is suggested by evolution.

Genetic Additions and Deletions

Evolution touts genetic additions as the evolutionary process advances, and it also talks of features being removed from creatures. Depending upon which evolutionist you hear speaking, you might hear conflicting views. I have heard people actually say that parts cannot evolve out and others say that part cannot evolve in or re-evolve in. But according to the overall idea of evolution this is utter nonsense. Either evolution allows adaptive diversification that can morph in to entirely new creatures, or it cannot. And if full-scope evolution is true then there are no barriers except for time.

Genetic additions and deletions is one of those moving targets that are difficult to pin down to dismantle and examine well enough to make any kind of substantive assessment due to the fact that one evolutionist will say "A" and the next will say "No it's B". If evolution is real, it will be able to add or subtract from the DNA and thus can modify itself as needed as it is being influenced by its surrounding environment.

If we were Created rather than having evolved, then we can expect that DNA can be both added to and subtracted from, but only within the confines of any broader kind. This means that an ape is not going to start growing wings at any point. And humans will never change much from what we look like now. This is true of all evidence dug up in the past and it will certainly hold true for the future, though only time will tell that story definitively. Our design is largely settled as far as all actual found physical evidence and the Bible indicate at this point in time.

Making Life In a Laboratory

We often hear of scientists in labs trying to "create life" as they fail miserably to do so. They create some interesting effects, but with trillions of cells for examples in each scientist's body to copy from, no one has yet been able to create a single living cell from scratch. If we come close to creating life, it is heralded as a grand accomplishment even though we failed. Government grants have even been either directly or indirectly given to labs to work on "creating life", and yet when the topic of abortion is discussed, then the millions and millions of cells in the "fetus" are not considered "life". This disingenuous societal dichotomy should be crushed by every scientist who seeks to read, alter, or affect DNA in any way. This is especially true of those who are trying to "create life."

The best laboratory in which to create life is the sanctuary laboratory of the womb of an actual woman. This occurs somewhere around three hundred times per minute all around the world with an impressive survival rate for the embryo to develop. There is a difference between a basic human cell and when the male sperm cell meets the female egg cell. Attempts at laboratory-created life fail in comparison with the intent and complexity of what occurs the moment conception occurs as that spark of life instantaneously begins after the male sperm enters the female egg.

Evolution ignores all of this and just assumes that life spontaneously arrived and the "sexes" spontaneously diverged, and then somehow decided to mate. Just in that short statement there are too many problems for it to be logically rectified.

When or how did the supposed evolutionary transition occur from being able to multiply from single replication, to requiring two-partner breeding replication? Was there an evolutionary period where the creature could both self-replicate *and* also breed with an opposite gendered mate? There are so many technical and logical problems with that idea that to tackle them all, along with the needed detail, is not practical in a single book.

There are a vast number of processes involved in cellular replications, and an even greater amount of processes involved in male/female reproduction. We can argue that any one single aspect of the processes has evolved, but to assume that all of the *required* processes were able to evolve to be dependent upon one another is really stretching the evolution theory beyond reason. What cannot be made up for with logic in the evolution theory is then compensated for with enormous amounts of elapsed time and ignorance in the form ignoring reality and pure logic.

We can try to make life in the laboratory, but in doing so we are proving our own incompetence in the matter while at the same time proving the superiority of the **I Am**. We may someday accomplish creating life, but I Am willing to bet my life on the fact that it will be a cheap imitation of cells that already exist using materials that already exist and processes that already exist and all with our immediate observation within our own bodies along with all of nature. In other words, we will be stealing the ideas from nature, copying those ideas and then claiming them as our own using our God-given Created intelligence to do so.

As a male's single sperm cell penetrates a female's single egg cell, the most amazing thing occurs at that penetration-conception-moment as a type of spark flashes. Following along the lines of the lightning-passing-information concept mentioned earlier, this is likely the same thing only on a microscopic level. We are somewhat logically forced to understand that this is the moment that, and how, the DNA information could be transferred from sperm to egg. Neither side of the debate should refute this because it appears that this is simply the way that the process is occurring in observation.

Our bodies are incredible machines with a vast array of functions, most of which we do not fully understand, and there are likely even more that we do not even know exist. The various known functions of the body that must work in harmony in order for us to live are many. And it is really very absurd to suggest that these harmonic functions that need each other for life to be sustained and propagated have evolved. Our designed bodies offer us warning signs of abuse, such as becoming overweight, or pain when you freeze your fingers, or indigestion from gluttony, etc. We can claim these are all a result of the evolution process, but because there are so many more such warning signs, it defies logic to attribute them all to evolution. Much like the warning signs we add to the dashboard of our cars, our warning signs are placed there deliberately so that we can make the needed corrections until the warning signs disappear. From an engineering perspective, there is no other option; most of these warning overload indicators are placed within us for our protection.

Chapter 15

Peer Review On the Same Page

Modern science is an interesting thing to observe when new ideas are injected into the existing dogma. A scientist will come up with a new theory or an adaption to an existing theory and then reveal it to the scientific community. On rare occasions the new idea will be embraced right away, but often it must be "peer reviewed" before it is widely accepted. This is a good thing because we should test other people's ideas before fully embracing them. In real science this is common, but with pop-science things are a bit different.

Pop-science's evolutionism is a bit of a cult, as is Young Earth Creation. In pop-science, often someone will inject a new aspect into the theory and then the rest of the cult immediately jumps on the evolution bandwagon to support the questionable addition. This becomes a dangerous trap to those who are vulnerable to cult-like beliefs. Because the pop-science talking heads are more marketers than they are scientists, they are expert at drawing people to them. Unfortunately, their audience sees them as "authorities" on the entirety of the evolution-versus-

Creation topic. Then when their audience chats with people about the debates, they do so without having reasoned through all of the cleverly presented data on their own, which then results in them further spreading the tainted information.

Observations versus Conclusion

An area most people struggle to grasp is the rationale used by the people proposing a theory. If you consider Darwin's work, you will find that there is a chasm between his *observations* and his *conclusions*. If you're not able to tell the difference between the two then you will struggle to recognize truth. Darwin wrote several books regarding his observations, evolution, and behaviors and emotions. In reading his books you will find many suppositions being cast in evolutionary stone. Statements along the lines of, [*if changes can occur in a given species and given enough time, then isn't it possible that a particular attribute could disappear entirely from a creature*]. We can suppose that this is possible, but we must realize that there are many other similar suppositions that he made that **all** of his subsequent conclusions are based upon. There are far too many *if-this-then-isn't-it-possible* statements in his work for it to be taken as credible.

There is nothing wrong with someone going through such mental exercises, and while he did devise some overreaching conclusions, it is not so much his problem as it is his cult following. It is his cult followers who fail to realize that his conclusions are his *conclusions* and they are not his *observations*.

Observations are the things that you actually witnessed. Anything beyond the actual things seen is speculative. So, we can see a humming bird and conclude that it has a longer beak than another hummingbird that we saw, but to devise anything beyond that is speculation even if that speculation is correct. Our speculation could be correct, but it could also be entirely wrong.

Darwin took his conclusions to more extreme levels with each book written as he slipped away from reality.

Darwin was studying to become a clergyman and the possibility that he owned at least one of those questionable Bible versions mentioned throughout this book runs high. Darwin had many great observations and did a fine job recording those observations and sharing them with the world. But when his Bible religion clashed with his real-world experience, it set him on a path that would deceive the world. He spoke to friends about his observations and the theories that he derived from those observations, but the ignorance of reality spewed by his friends caused Darwin to dig his heels in to stand his ground, ultimately causing him to stretch the truth so far that it eventually became a lie.

Observation is what you see and experience with no "facts" added, *conclusions* are what you think about what you saw, which can be highly creative. It is in what you think about what you saw where new so-called "facts" are often inserted. But those are actually not true facts and are, instead, *suppositions*. This is even true of the book you are reading right now, information is analyzed and then using current natural experiences, logic, and reason we can then make some assumptions. The question that we must then ask ourselves is, is there any evidence that contradicts our theory? If there is, then we truly need to take a look at that evidence to see if it nullifies any or all of our theory. That is to say, are there any testable observations that will nullify a theory?

Simply pitting one far-reaching theory against another does nothing to prove or disprove either theory, but actual observations are very powerful in this regard. So, saying that evolution is not real is simply not true. But the scope or limits of evolutionary-change is likely confined within the "kinds" stated in the Bible, not because the Bible says so, but rather because this is what we actually see in nature, and it is also what we see due to the gaps in the evolutionary tree of life. Evolution struggles to bridge those gaps between Biblical "kinds" and has not yet legitimately done so.

Know It All

The worst beating that takes place in the evolution-versus-Creation debate occurs in colleges when dedicated Christian students attempt to challenge their college professors on the topic of the existence of God or on the evolution-versus-Creation issue. These beat-downs are usually very brutal and have turned many students into atheists. This occurs because they have been so built up in their false-beliefs and never had the opportunity to experience any pushback of their incorrect interpretation of the Biblical Creation in Genesis One. And, I might add, that their interpretation of the Bible is actually almost always someone else's interpretation of the Bible, and they are simply regurgitating it much the way the evolutionists regurgitate the religion of Darwinism. The only difference between the two sides is that the Christians are typically kinder about it. Both the professor and the student behave as if they know it all, when it is more than likely that both have many errors in their own position.

The one point that I cannot stress enough for both sides of the topic is to get a hold of an authoritative Bible and understand it well before taking part in such debates. This is discussed in *Understanding The Bible - The Bible How-To Manual* AND *The Things We Don't See, Understanding The Church - Upon This Rock I Will Build My Church* and in the various volumes of *The Science Of God*. While we act like we know everything, the truth is that we simply do not. A fair amount of evolution and Creation theory is largely speculative.

We don't even know how or why the bones form in a developing embryo, or how the blood forms for that matter. How can we expect to know that our way is *the* way? Our quest in this debate should always be for truth, rather than always trying to further our own agenda. If you consider the full-scope evolution theory of amoeba to primate to man, or the ideas presented in the various *The Science Of God* Volumes, don't just believe what you are told, use your God-given logic and test the ideas to see if they actually work in reality. Look for flaws that are provably incorrect, or are over-reaching and are not able to be explicitly proven, and then ask yourself if those theories are over-reaching and unrealistic, or are they realistically plausible and even likely.

Progression of Evolution

The theory of evolution has evolved more than the animals and people that the theory professes to have evolved. A big contributor of evolution is the ignorant idea that animals are stupid and have no sense. This view sparked curiosity in Darwin which caused the close observation of animals, causing Darwin to finally realize that animals communicate and have intent and can make decisions.

When we drop our ignorance and pay attention, we can then see the brilliance of the animals all around us. The foolish assumption that they don't have feelings or are not smart is due to our foolish human arrogance. It is the same arrogance that causes racism when the followers of evolutionism believe that they are superior to other people, which is something that ran rampant in colleges around the turn of the twentieth century.

When we assume that animals have no sense about them, then we fool ourselves into believing other lies as we begin to realize that animals can also express some amount of emotions or decision making as Darwin eventually finally realized. But some of the responsibility of his ignorance in that matter is partly due to his religious beliefs.

When you understand that the animals were Created as they were "brought forth" by the "earth" and "water", and that the **I Am** infused instruction into them, you truly should have no logical choice other than to assume that animals are "smart" within their own realm. Just because we humans have dominion over the animals does not make animals stupid. When we see these God-given aspects of animals, we tend to see the animals as more human-like, thus causing some of us to believe we evolved and that the attributes are both inherited and learned. What we fail to see is that designers use common parts for multiple purposes and when we design things, then that which we design will share attributes. For instance, when singers write and sing songs, their work is typically recognizable as *their* work. Thus, God's work is going to be recognizable as God's work, or, at minimum, you will see traits that are common amongst the various creatures because they were Created by a singular Creator whose signature is the common attributes and structural components that we see in such creatures.

Fins work, so fish have fins. Feathers work, so birds have feathers. This does not mean that there are no other ways for creatures to fly or swim, it's just that those with the most efficient attributes will thrive in the environment for which they were intended to dwell.

But man is unique. Look all around the world and at all of the animals and you will see animals thrive where they are best suited. But with humans, we can thrive anywhere, and we do extremely well in moderate climates. Most animals generally do not thrive just anywhere unless humans are there to help them. Man is very obviously special and no obfuscating of facts will change that simple truth. I encourage people to examine evolution, along with logical Biblical Creation while using an authoritative Bible. Doing so will prove *your* worth. Do some research to see what you find: Has the theory of evolution itself evolved since its inception? Is it proven to be immutable fact? Is logical Creation possible? Think for yourself in these matters.

What is "Evolution"?

Evolution is a theory of progressive successive minor changes that allows for an ancestral creature to morph through its lineage into an entirely new creature tens of thousands and millions of years later. Yet while evolution as just stated is theoretically possible, we have to remember that "evolution", or changes over time, actually occur very quickly in reality.

Consider when people crossbreed animals, that in a single generation they create an entirely different creature that is within the same Biblical designated greater kind, and this process can be repeated to evolve attributes out of the animal and sometimes into the animal. This process is common with dogs and cats. So with that rapid of massive change in a single generation, you would think that in ten or twenty generations we would be able to crossbreed the animals, such as a dog, into something that resembles a goat, yet we don't see anything anywhere near departing from the fundamental kind of even a dog, let alone departing from the specifically stated Bible "kinds" that are far more broad than are designations such as a "dog".

In animals, which man was given dominion over, we can quickly alter the design of the dog, yet the dog remains a dog. But with man, that design has historically shown very little variance according to all serious fossil finds. Sometimes people will point out extreme birth anomalies as some sort of evolutionary indicator, but people with such anomalies often do not survive long, and if they do, they seldom reproduce, thus maintaining well-structured "man". Minor defects are typically corrected by DNA in subsequent generations.

The evidence is pretty clear that man did not evolve from anything and the emergence of man was abrupt, and complete in the geological record, although, the geological record is also read in error, but that's for *The Science Of God Volume 5 - Boats, Floods, and Noah - The Deluge* to explain.

Pop-evolution, as promoted, is change over time and it is without limits. Real evolution is more of an adaptive function in the design of the creatures as a form of survival for the various Created "kinds". We see a lot of this change in animals, but no substantial change in "man" has ever been witnessed in any archeological dig, ancient writing, or in current reality.

It is important for us to get our heads straight in these matters. Because when we believe we evolved, then we tend to not realize the deliberate warning indicators that have been designed into us, including our confusion when we foul our logic. If you ever get confused or at all foggy regarding the evolution-versus-Creation debate, take note that it is a Created indicator that you have compromised your logic and need to step back, take a breath, and then re-evaluate you current understanding and position.

Chapter 16

Ten-Thousand Scientists Can't Be Wrong

Numbers do not make something correct. When there is a "scientific consensus" regarding evolution, it does not make their understanding correct. Full-scope evolution tends to be the consensus in the science community; however, there are many scientists who remain quiet on the matter for fear of losing their position or job. This would not be that case if people understood the Genesis interpretation dilemma that has been created by the late post-Reformation Bibles. The problem is that people simply do not realize that this Biblical interpretation issue exists because they were taught that the Bible is "inerrant" and that *their* Bible is *the* Bible.

We need to have the courage to stand against error no matter how many people have chosen to believe in that error. But we had better be able to back up our own claims and our refuting of their claims. If we make our case, they still might not accept or possibly even understand our own position, but if we fail to convey the truth to them, then we are going to be partly to blame for their demise on judgement day. This connects to the whole point of "The Church" as mentioned in the book *Understanding The Church - Upon This Rock I Will Build My Church*.

Numbers in consensus do not make those numbers correct. Only Truth is correct, and if we fail to embrace Truth, then we will follow an erred pack of scientists to our own destruction. I realize that this sounds a bit harsh, but the Bible speaks clearly about false teachers, which includes anyone who is teaching incorrect information in the form of illogical theories and is drawing people away from God. Doing so is, as a technicality, being a "false teacher". But this is also true of those who teach six twenty-four-hour day Creation. Christ said "It were better for him, that a millstone were hanged about his neck, and he cast into the sea, than that he should scandalize one of these little ones." Make sure that what you teach is accurate, or at minimum clearly stated as *only* a theory or viable possibility.

Being Specific About Defining Terms

The scientific community is often very specific about defining terms and insisting others abide by those terms, such as is mentioned in the *Phylogeny* section in Chapter 9. But often they do not adhere to the same standard that they hold others to. In the real science world this is not a problem, but in the pop-science world it tends to be the norm. There is a sort of blindness that has descended upon the world that has pitted the two sides in the evolution-versus-Creation debate against one another to a point where they get trapped in the process of rejecting the terminology and vernacular in each other's beliefs.

If the evolutionists would stop rejecting the Bible altogether, and if the six-day Creationists would stop rejecting any notion of evolution and not be afraid of the word "evolution", and if both sides would be more specific about terminology used in the Biblical account of Creation given in the major authoritative Bibles, then we could discover new things. But stuck we will remain in the muck and mire of this debate until we all awaken from our ignorant slumber.

Standards are key to everything! Science, which is to say splitting and dividing or grouping things in an organized manner, is the first action of Creation stated in the Bible. To its credit evolution-science tries to group things, but it ends up creating an endless list of groups in effort to be more specific.

We need to regroup with regard to this debate and consider **All** of the data from all sides. We need standards in the definitions of "species" and of "kind". We also need to stop trying to make the Bible or the fossil record say what *we* want it to say, and instead we need to read the data for what it is. If you step back from this debate to zoom out and get a birds-eye view, you will very quickly see how frequently ideas are inserted into the actual data and then those insertions become the foundation for the belief when the actual data has often been misunderstood.

Getting Published

College causes a problem with its un-doctrination through indoctrination. And when adding to the quest for recognition, the un-doctrination problem quickly gets worse. As students and professors go head-to-head on the Creation and Creator subjects, the professors often win simply by default due to their status position and authority over the students. If they don't want to be challenged, they can downgrade a student into submission. After the students submit to the professor's position and philosophies, then they will be taught how to write papers for publication.

Getting published is a cherished accomplishment that can inflate the ego of the student-scientist. Even after graduation, getting published is critically important to their future career plans. There are standards for getting published in scientific journals that must be abided by, and one of those unspoken standards is to not mention the Bible or Biblical Creation in any positive or major way. If someone was to incorporate a position that is favorable to the idea of deliberate Creation of "man", or guided Creation in general, they reduce the possibility of getting published by nearly one-hundred percent. So Biblical Creation is essentially not allowed in the discussion at the scientific publication level.

If you were fortunate enough to have gotten published and wish to use that as a credential for getting a faculty job at a university around the turn of the century, you were unlikely to do so if promoting Biblical Creation in that article. In all fairness, many of those who push Biblical Creation offer only the nonsensical six-day theory taken from the post-Reformation Bibles. But even so, the scientific community has been so indoctrinated that defenses immediately are put up upon hearing the words *God*, *Bible*, *Flood*, or *Creation*. So, anyone who is using logic and commonsense to understand and disseminate the Creation text is going to have to be very articulate about how they make their case. The same level of care need not be taken when writing on evolution because the many erred presuppositions are so deeply embedded in the mindset of the evolution adherents, of which there are many in the colleges. To get published you only need to have something new to say about evolution to the scientific community, or even only a new way to say the same old thing, and then format and submit it properly and you then have a very good chance of getting published.

Creating Theories

Since evolution has been quite thoroughly addressed in publications over the years, it is a bit of a task to say something that has not already been stated in previous publications. Some things can be reiterated by subsequent writers because old information becomes new again to each new generation. But often people will grasp at straws to invent a new angle so that they can get published. Getting published at one time was about revealing new information to the world with facts and figures, but since the basics have been covered long ago, scientific writers generally must create new theories to garner enough attention to be worthy of publication.

With the proliferation of technology in the world of publishing, we are exposed to an ever-increasing onslaught of material to choose from, thus making the publications market very competitive. During the late twentieth century this problem increased tremendously causing increasingly outrageous articles to be accepted for publication, which has not yet stopped. In fact, the problem became worse with the advent of "digital online publishing".

What few people seem to grasp is that all magazines, papers, and radio and TV shows, etc. all are there to make money. They too are trying to earn a living. Their revenue and paychecks come from the advertisers that run commercials in those venues. The advertisers often have no concern regarding the subject matter of the show as long as people see the advertisement. This means that the people who run these forms of publication and entertainment do not need to concern themselves with the validity of the content. This is true of nearly all publications–including some scientific journals.

The media market is oversaturated with publications and is hungry for content. The more outrageous the content is, then the more likely it is to grab the readers' attention from shear outrage alone. So as long as a scientific article crosses all of the "t"s and dots all of the "i"s and follows the fundamental guidelines for publications, it is likely to be accepted for publication if it has any bit of new information in it. In other words, it only needs to be presented well and be a premise or theory, but it does not have to be true or possible–nearly anything goes. This brought about articles being created for the sake of articles being created just so someone can get "published" and/or make a few bucks. This is why there is so much garbage published these days in the digital culture.

Our Ingenuity

Interestingly, our ability to imagine and fabricate well formatted lies and deception for publication, is an abuse of our obvious Created nature. While it is an abuse of our **I Am** type nature to lie, that nefarious creativity also displays our ingenuity regarding the creative writing that is used to invent such fantastical evolutionary tales. The difference between our misuse of ingenuity versus God is that you can test Creation in any way possible and it will always pass scientific processes of logic and truth when it is properly tested, where our evolutionary lies simply can never pass such tests when allowed to be properly and truthfully tested.

Chapter 17

Our Attempts to Share

In listening to preachers who talk about the Creation of "man", it is typically presented through pie-in-the-sky rose-colored glasses where everything was done to perfection and God had it all planned out in advance. And when it comes to the initial points of Creation, we tend to group it as one big Creation action, yet that is not what the Genesis text indicates. Genesis doesn't say I am going to do this and then this and then that. It says the God did an action and then summed up that action as completed signifying that "it was good".

God first Created the basics and then Created light using those basics and separated light and dark and waters above from waters below and Created plants and Earth and animals and then God changed things up a bit and explicitly stopped to make a Creation in the image of the **I Am**. It is pretty obvious that God wanted someone like-minded to share Creation with. This nature is within Father Adam and Mother Eve and their offspring as we are all Created "in the image of God", and we all have the Breath of God in us.

We all have a deep desire to share ourselves and our work; that is until we are assaulted by the world for doing so. Consider any child who does something that gets attention; for instance, a work of art that they created for Mommy and Daddy that they will proudly show to you. They want to share. Sharing is natural in us, but when others try to take what is rightfully ours, we then withdraw to protect ourselves. In the Garden of Eden when Satan deceived Adam and Eve, it was an attempt to take from them causing them to hide from God in fear, which is exactly what we typically do when we are wrongfully robbed of our rightful place in sharing ourselves with God and our fellow man.

Even if you don't believe that a Creator exists you will still exhibit a desire to share as is made evident with people's attempts to be heard and published. It doesn't matter if what we are trying to share is right or wrong, we still all generally have the desire to share.

Becoming Civil

The evolutionary tale leads us to believe that "ugh-ugh" cavemen once roamed the Earth and that they were mostly without sense, but these imaginative tales are nullified by reality.

The idea is that since we allegedly evolved, we began without any knowledge and slowly acquired knowledge and through that we eventually assembled coherent communication that developed into speech. In this view, our ape-like ancestors were potentially quite savage and did not have our current-day rational nature.

Some of the Aborigines were cannibals and Darwin observed some cannibal Aborigines who were made civil by the missionaries. So we have to question if these Aboriginal people of relatively recent times were isolated and did not evolve with the rest of humanity who they happen to resemble quite astonishingly, or is it possible that they de-evolved?

Keep in mind that it is very evident that in nearly every culture on Earth, people all around the world have retained histories going back thousands of years, illustrating that they had fairly sophisticated societies.

Did Darwin's Aborigines of the eighteen-hundreds de-evolve or were they left behind? This division is not a structural physical difference; rather it is a cultural difference. Darwin alleges that some of the Aborigines that he encountered on his travels were cannibals with unseemly personal habits and hygiene. This lack of sophistication on the part of Darwin's Aborigines flies in the face of evolution and tends to lend to the idea of de-evolution.

In the Bible it speaks of "fornication", and we wrongly assume that this only means having sexual relations before marriage. And while that is our cultural understanding, it goes much further than that. If you read the Bible closely and pay attention to what is occurring when somethings are talked about in the Bible, you will quickly see that while fornication appears to have included some sort of sexual deviancy, it also included sacrificing objects and food to idols, including sacrificing people, and then consuming that which was sacrificed. The practice of fornication was sin that defied everything about God and Creation, and in many cases, it was cannibalism.

It is clear when looking at history that the societies that embraced God thrived, and all those that turned away eventually failed or were severely stunted. We have no evidence of human evolution, but we have abundant evidence of human social de-evolution and deviancy. Yet through it all, man remains man, fallen but still in the form of man. We do not become civil through eons of evolution as Darwin and so many of his followers imagine, rather we become uncivilized when we depart from the **I Am**.

Bible Abroad

Darwin observed that the sharing or spreading of Christianity abroad had resulted in civilizing tribes of natives, and yet he gravitated toward denying God and denying the deliberate Creation of man being made in the image of the **I Am**. His observations of "natives" make it quite clear that people hunger for truth and will quickly adopt that truth when it is actually offered to them. There is no evolution needed to bring people to this higher level of understanding. The Aboriginal people of Darwin's time are of Created man, and they too are made in the image of God who is The **I Am**, and therefore they have the capacity to quickly learn truth when it is actually finally offered to them.

Cultural forces are what limit us in our understanding. It is only when someone is brave enough to speak truth that culture can advance. Darwin did that in the beginning of his adventures, but through stiff-neckedness he and his intellectual foes each stood their ground and fell ever-deeper into their own erred conjecture. We have centuries of proof that when people hear and then properly follow the information conveyed in the Bible, the societies in which they live improve and become more civil, but only if they properly follow the Biblical ways. However, when they depart from the Bible, then those societies slowly slip away as was made clearly evident in the United Stated of America around the turn of the twentieth century.

Same as 3500 Years Ago

Darwin wrote of natives cutting themselves upon the death of others, which is something that the Bible speaks against. We are not to mark ourselves for the dead. The interesting thing about this is not that the Bible prohibits it, but rather that it was prohibited because people were doing it way back then. In fact, all of the guidelines in the Bible are there because people were doing the very things that the guidelines were put in place to prohibit. History does repeat itself as archeology, Darwin's observations, and The Bible all clearly prove. Beware, because it will come to your community if your community does not stay vigilant in such matters—if it is not already there.

Chapter 18

Believe What You Will

At first-breath, man was instilled with the desire to believe. Belief is one of those misunderstood words that are all too common. In the Gospels, Jesus talks about children believing and says "Truly, I say to you, unless you turn and become like children, you will never enter the kingdom of heaven. Whoever humbles himself like this child is the greatest in the kingdom of heaven."

"Belief" is our capacity to accept something as true. "Belief" is the *level* of trust or love we have toward an idea. We can believe whatever we choose to believe, but often what we choose to believe is not what we really believe deep within ourself.

Filling Our Desire

All of man has a deep desire to know things. We might want to know different things, but we all want to know things. For instance, Adam and Eve wanted to know the difference between good and evil, and they did succeed. It cost them a great deal, but they accomplished the task. Our curiosity is a part of our being "made in the image". God was obviously curious and investigated, and then Created based upon that curiosity. *Curiosity*, which is to *wonder*, is one of the base concepts instilled in us at our Creation. It is in itself the ability to desire, or maybe better put, desire is the ability to be curious.

When we have gaps in our understanding, we have a natural inclination to go on a quest to fill those gaps. That is exactly what the **I Am** wants us to do, not in spite of God as is often thought and done, but rather to know and understand God better. When we are Creating, then the "gap" is in the question, "What will happen if I do this?"

Arrogant Statements

The evolution side of this debate tends to be a bit more arrogant in their presentations, but that's only because the Creation side is typically Christian and is taught to be humble. You will hear some of the talking heads of pop-evolution-science make arrogant statements that some things in the body are a "poorly designed", thus if a designer made them, then "that designer was stupid". But this ignorant approach fails to address one glaringly obvious oversight: These so-called "poor designs" have been successfully serving humanity very well for many thousands of years as far as history and all evidence shows.

Evolutionism is at somewhat of a disadvantage, in that their assumption is that everything evolved, thus any bad designs would have to be considered evolution failures in respect to the "poor designs" comment. But the true disadvantage that occurs with evolutionism is that the assumption is made that there is no designer and thus there is no intention or reason that things are as they are. If you make an assumption about the design of man by using evolutionary philosophies, then of course, in your mind, man is not actually "designed" but rather is just evolved at random according to the surrounding environment, and therefore you will ask no additional questions.

However, if you make the assumption that the Creator/God does exist and intentionally infused specific design information into Adam in order to Create him, then you will question *why* something is so. You will wonder upon the reasons for *why* something functions as it does. And without that information gap, we typically will look no further.

Our desire to learn and to know things is partly a *how* situation, but more so, our desire to learn is a *why* situation. There's an enormous difference between asking *how* something occurs, as opposed to asking *why* something occurs. Evolution has no concern for asking *why*. People might unknowingly state a question in the *why* format, however, when no form of intelligence is present, *why* is technically out of the question and it is a *how* situation. If a student asks the teacher *why* a cell divides, often the question is actually meant to be how does that occur, which is almost always the manner in which the teacher will answer the student's question.

Science's quest is more in line with *how* God did things, rather than *why* God did things a certain way. So, arrogantly implying that if God exists, then God is a poor designer foolishly fails to ask *why* something was done a certain way.

We have been imbued with curiosity and reason by the **I Am**. It is time that we all start to take advantage of that gift and recognize that *why* questions need to be properly developed and then asked. Asking *why* should enhance curiosity in man and cause us to seek ever more answers. But when the *why* is stripped from our imagination, then only *how* exists for us and we then lose interest. On the evil side of things, forbidding *why* by only asking *how*, quells curiosity and is a directive of evil so that we do not become closer to God. If you test this or look for this you will recognize it showing through in most of the evolution arguments.

A Fishy Story - Hook, Line, and Sinker

Darwin's clever statements lure the reader to suppose many speculations as true and then proceeds to base subsequent suppositions on the previous speculations, which he then extrapolates to arrive at his outlandish conclusions of mere supposed possibilities and presents them as pseudo-"facts". This is not implying that all of his theories are absolutely out of the question, but rather that they are not absolute, and while some points could be spot-on, others are completely in obvious and absolute error.

Evolution has many such false premises. One of them is the fish to human origin theory were they look at a fetus in the first couple of weeks of development and see "gill-like" structures on the fetus' neck. But this very assumptive view overlooks the basics of the progression and fundamental structure that most life has. With this ridiculous rationale we can assert that we evolved from any creature because when the first cell becomes an embryo, then **all** life looks pretty much the same.

As the cells multiply by dividing and increase in structure there are certain key aspects of life that need to occur first, so most creatures will have somewhat similar forms at the earliest stages of life. You can think of this in terms of resolution and of the dots that make up a digital picture. The fewer the dots the more obscure the image is going to be. As the dots increase in quantity, the picture becomes more refined and the distinct features become more apparent. While this is not a perfect analogy, it illustrates that with more and more cellular resolution and with size of the cluster of cells, each feature can begin to show its distinct nature. These early development similarities are not because we evolved, they are because of the obvious design used in the greater part of living creatures.

Micro Evolution

In discussions surrounding evolution, you will encounter terms such as micro- and macro-evolution. Macro-evolution is the full-scoped evolution that we have been discussing throughout this book, which goes from amoeba to primate to man over long periods of time. Of course, there are theoretically more transitions in the macro-evolution theory than in micro-evolution, but that sums up the macro-evolution theory.

Micro-evolution, on the other hand, is changes *within* a "species" or "kind". This is the common and obvious adaptive evolution that we can witness while we live and breathe because it can occur in only a single generation with every generation of offspring. Micro-evolution is a term some Creation supporters use when referring to "adaption" of various environments for a given species. Even a human child is the result of micro-evolution because while they are of the same kind as their parents, that child is distinctly different, and parts of the attributes of that child will be passed on to its own offspring when that time comes. This is undeniably true, demonstratable, repeatable, observable, and verifiable, using testable facts, and this aspect of micro-evolution has never shown anything that steps outside of the parameters constituting "man" as set forth in Genesis One on Day Six. This is clearly illustrated by the simple fact that humans are all human with no deviation found whatsoever.

Is a Fetus Really "Just a Clump of Cells"?

In our modern era, we are actually not so "modern". In fact, if you take the time to study the act of "fornication" as noted in the Bible, you will quickly see that people were sacrificing their children to idols and other evil entities. This is similar to abortion where we humans will kill and sacrifice the infant within the womb for the sake of our own personal comfort and greed. But does this matter? A fetus is just a clump of cells after all, isn't it?

Let's take a look at what happened when *you* were conceived. Truth be told, most children are conceived before any sperm cell actually gets near any egg cell. Most offspring are conceived first in the mind of the woman and then when she marries, the child is then typically conceived in the mind of the man as well. Now of course, there are those who claim ignorance on these matters and just have "sex" which then produces a baby, but even then, the child is conceived before the act ever occurs, but then it is as a side effect rather than as the primary goal of that particular instance of the fertilization activity.

When we first imagine something and then ponder it, that then is when something is conceived. And in general, without that premeditation, most of us would not be here today because either we ourselves or an ancestor would never have been conceived or existed. So clearly, we are generally conceived first in the hearts and minds of our parents.

But keeping things a bit more tangible here, let's see what happens when the sperm cell meets the egg cell. The male emission from a single romantic encounter has upwards of fifty million of the little guys swimming about in the semen. And when properly deposited, these little guys rush towards the female egg in a manner similar to boys rushing to pretty girls. When the male cells are romancing the female egg, then typically only one male cell will be accepted into the female egg. When this takes place then that spark of life mentioned in an earlier section occurs. Regardless of your view on evolution or Creation, this is the point where some DNA data is being added to the female egg via electrical impulse. It is at this point that within about a single day the newly fertilized egg begins to multiply by dividing.

Within only a few days this highly detailed data becomes a "clump" of cells. This newly formed "clump" was *you*, and it contained roughly twenty new cells at that moment. Each cell carried your entire DNA message and somehow got its very own personal instructions as to what its function was and what part of your body it would become. Your head and spine started forming immediately as a part of the "clump" because they are the core of your body. In only about seven days your blood began to form, and in less than three weeks you had a detectable heartbeat. In as little as three weeks' time, you had lifeblood flowing through your veins.

By the fourth week you had developed noticeable formations that eventually became your hands and feet. By the eighth week you were just over one inch tall. And by ten weeks you were formed with all of your organs, but you were only about two inches tall then. From that point onward, you grew and relaxed in the comfort of the bath of your Mother's womb until you no longer fit and you wanted out.

Regardless of what stage of your development, through every single stage of your development, **you** were **you**. You were Created in the image of your Mother and Father who are Created in the Image of God. And since you are a clone of your parents, you therefore are also Created in the Image of God.

All of the systems put in place by the Creator are miraculously replicated every time a baby is Created "in the image of". These intricate and interdependent systems allowed you to live and breathe just as they do right now as you are reading this. This replication procedure of "man" is inarguably true, as we witness this on a daily basis with babies being born who resemble the mother and father parents who actually contributed life material to the baby's Creation.

For about nine months as you held your breath and basked in the bath of your Mother's womb, you came to that higher state of being and were ready to exit your Mother's womb. Then your body prepared itself for your departure from the safe and peaceful comfort of your Mother through your door to the world. Almost immediately upon leaving the loving comfort of your Mother, you took your very first breath and then began to cry. You were now no longer together as one with your Mother, you were now delivered into the world as a unique and fully independent Creation.

Within your Mother, your unique cells multiplied by dividing until your body was fully formed. At this point your newly delivered body vessel was examined for a moment and then dabbed off with a towel to clean you up a bit and you were then handed over to your mother who held you close as you nuzzled for the nipple on her breast. You were committed to this task as was your mother and those around you, and by some not-fully-understood amazing instinct, you tried guiding yourself to your Mother's breast that would provide nutrition for you to live as you began to, or at least tried to, nurse from her breast no matter who was holding you.

Your system was built the same as your parents and you were now a fully independently functioning copy of your parents. At some point you opened your eyes and began to gaze into the eyes of your Mother or Father. As your Mother held and kissed you, the two of you shared a special bond and your desire for each other grew and was embedded and retained in both you and your Mother's heart and mind.

As you suckled milk from your Mother's breast, your body systems and all of your organs began to process the milk and utilize it to help you grow. Now at this point you are only a few hours old after being borne into this world. You could hear, you could see, you could taste and smell, you could feel touch. You could sense comfort and love, and even fear. You were fully man and fully made in the Image of God—You were now *you*! And nearly all of this was also true many months before you ever left the comfort of your Mother's womb. It's just that then you were so small and vulnerable that even with great care you would have struggled somewhat to stay alive, so you chose the comfort within your Mother until you were ready for the world

When you were so young, you had not yet learned of the sacrifices your Mother made to give you life. You did not know how or why she used her blood to keep you alive as her blood exchanged nutrients and oxygen with your blood in the placenta so that you could live. Her body's promise to you was fulfilled through your shared placenta that she produced for you to keep you alive and healthy. That placenta filtered out most contaminations that your Mother might have been exposed to in the world, and all as a promise to keep you safe and healthy. You are now "man" Created in nearly the identical way that Mother Eve was Created.

We can believe whatever we want about life, but truth is not something that our belief will ever be able to change. You are uniquely Created in the image of your parents *and* in the Image of God. To entertain the idea that all of that and all of the interdependent systems evolved without any intelligible guidance whatsoever is really quite absurd. You are Created, rather than having evolved. Just as your parents used intelligence to conceive you, even far more so did God use intelligence and wisdom to Create you. You are conceived just as God's thoughts are conceived.

Your Soul began in the thoughts of your parents, and it came to existence the moment your Mother's egg cell was penetrated by your Father's sperm cell. However, just as the self-Created **I Am** had to come aware, so, too, did you have to come aware. So, your initial self-awareness was immediate at the point of fertilization, allowing you to "become" much faster than the I Am Creator did. From that point on you had to learn the key concepts on your own via experiences from your senses. Your Soul, that is to say **You**, from the moment you *became*, had to learn and *become*, similar to how the **I Am** Creator *became*–step by step grasping each basic concept on which to build your understanding.

A soul is something that is built bit by bit and never stops being built, at least while you are alive in tangible Earthly form. A strong Soul has understanding, and understanding is a never-ending ongoing process of awareness. Always seek to understand in order to increase your Soul because it will either be built with Truth or it will be built with lies.

Chapter 19

Preserved for Safe Keeping

At our Creation, we were given the wonderful gift of retention. Without this primary conceptual function of retaining things, nothing today would exist with or without a Creator. Retention is stability, and without stability the atoms could do all sorts of unexpected activity. Without stability we have nothing worthwhile. This is true of our thoughts as well. We have been given the ability to retain and remember things. Without our ability to abundantly retain we would be somewhat animal-like.

What we choose to remember will determine our path in life. If we remember lies or incorrect information, we eventually will make wrong choices surrounding that which we believe wrongly about. For instance, if we foolishly accept that the Bible is mere fairytales, then we will assume that the Genesis Creation account is fake and that the **I Am** Creator God does not exist, causing us to behave in such manner so as to not believe that Christ came to redeem us all from Adam and Eve's error of stealing from the tree of knowledge of good and evil. This comes down to the fact that if Christians are correct, then we will face a tormented eternity of conscious awareness without God where there is "weeping and gnashing of teeth". We have been given this gift of the ability to abundantly retain truth, but often we use it to retain lies.

The Earth also has a wonderful retention system, except the Earth is a bit more as-a-matter-of-fact about it. The Earth has entombed many thousands of years of history in rock. The term "Where is it written in stone?" is coined because of the Ten Commandments that were written in stone. But Earth also has, over the years, written the "geological record" in stone for us to peruse at our leisure. The difference with the Earth's ability to retain information, versus our ability to retain information, is that the Earth cannot lie, but we sure can. We can take the Earth's memory record of past creatures and man's past civilizations and then imagine all sorts of outlandish scenarios as is done with evolutionism.

The thing that makes full-scope evolution so convincing is that it is not **all** lies and misinformation. Evolution has many great observations, much like Darwin found. However, just like Darwin, the father of evolution, the lies and misinformation come in the form of *conclusions* that are built upon the sand foundation of imaginative supposition, which then are all presented as "fact". The most obvious of which is the issue of time.

Separation of Ages

If we assume that the Genesis Creation account occurred over six twenty-four-hour days as is mistakenly implied in some late post-Reformation translations of the Bible, then we immediately are at odds with reality *and* with science. However, when we understand that there are no time limits stated in Genesis, then Genesis One and Two start to make a lot of scientific sense.

Interestingly, man seems to have sprung up very quickly on Earth, and all research and evidence of evolution agrees with that.

This sudden arrival of man supports the Bible's sixth day statement about man having been made after the animals. When we look at the geological record, it shouts out everything just as is laid out in Genesis. But lest anyone misunderstand this, the geological record has some not so secret secrets that are discussed in *The Science Of God Volume 5 - Boats, Floods, and Noah - The Deluge* regarding the flood of Noah's time.

There are distinct ages that appear to have clear separation and they are as follows: Before any life, After plants, After animals, and After man. Each of these is clearly laid out in Genesis, but it was not done in twenty-four-hour days as some people have been duped into believing. The ages are clearly defined in the Bible, but in the fossil record things are a bit different.

Foot Prints Written In Stone

Evolution proponents often encounter dinosaur footprints in their search for fossil relics, and when those footprints are found those prints are immediately gauged to at least sixty-five million years old, because that is how long ago dinosaurs are believed by evolutionists to have become extinct. When footprints of man are found they are typically gauged to be much less than two-million years old and often just tens of thousands of years old because that is the alleged beginning of man.

There are allegedly footprints of man found in the same piece of rock that dinosaur footprints are found in. If this is true, then that would completely upset the idea of the separation of ages between dinosaurs and man. But, even if such evidence does not exist or has been fraudulently entered into the data stream, the fact still remains that footprints of man clearly exist on the same plain or layer as dinosaur footprints, but are miles apart. These footprints are not dug up from scores of meters or feet below the surface; no, most footprints are found on the surface while people today are walking around, hiking, or maybe digging in very shallow archeological digs of only a few feet in depth.

The problem with this is that the *ten-thousand-scientists-cannot-be-wrong* scenarios are so strong in the science community that the analysis of data is always tainted to lean toward the suggested consensus of geological-evolutionary ages. It would be good if both sides could overcome their biases, but since that is not likely to happen anytime soon, it is up to **you** to view that data and use your own true logic to ascertain the actual Truth.

There is such a wide array of geological upheaval and erosion that it can be somewhat complicated to know when a layer of sediment was actually deposited. But what we can tell is that each layer was typically laid down quickly. For instance, when we have a layer of limestone that has dinosaur footprints in it that are several inches deep and that layer is only about twelve or less inches thick, then we can safely assume that the layer was deposited in a matter of months or days or even only hours. Further, judging from what we see in our general life-experience regarding substances, if a layer of material is several feet thick and it has dinosaur footprints in it that are several inches deep and the layer is contiguous with no separation layers and the material is consistent throughout the entirety of the layer, then we can be pretty sure that it was deposited in a fairly short period of time.

The question about these treasured footprints is that, if two sets of prints are found in similar rock but are many miles apart, and one set is from dinosaurs and one set is from man, then how do we know that one is over sixty-five-million years old and the other is less than two-million years old? Even radiometrically dating the rock in some manner will be influenced by the knowledge of what is being dated.

We are running out of time for total honesty as more data is collected and compiled into databases available to the public that are built on the assumption of the evolutionary ages typically used. Radiometric dating is more of an art-form than it is a science, as is proven when newly formed rock is dated as tens of thousands of years old when the rocks formed are actually known to be less than only decades old, as is the case with the Mt Saint Helens eruption. If enough analysis data is collected from radiometric testing, then eventually labs will be able to have a fairly good idea where a piece of stone is from using the collected data. This will make blind-tests and total honesty become a thing of the past.

Anyone who builds computer systems and data analysis equipment understands that calibration is key when calculating any data from any data acquisition devices. And we get to calibrate that to suit our needs. For instance, a scale will have "load-cells" that need to be calibrated via the software so that when a pound or a kilogram is placed on the scale, then the scale display needs to show the appropriate value. But that value can be any value we want to make it. And as long as it matches other scales, then everyone is content. Similarly, with the various radiometric dating methods, calibration is key and the values that are established in the software are decided by the equipment and software designers. While radiometric dating is a bit more complex than simple load-cells from a digital scale, the calibration issue remains exactly the same.

Macro-Evolution

The separation of ages that is implied in the theory of evolution is what allows evolution's theory of macro-evolution, which is to say full-scope evolution, to be allowed into consideration. If the ages are removed and time is shortened, then macro-evolution falls completely apart.

Micro-evolution is accepted by both sides of the debate, but is generally known as "micro-evolution" by one side and "adaption" by the other side, but that is nothing more than a semantics issue.

Macro-evolution, on the other hand, is solely dependent upon time. If you take long periods of time away from macro-evolution, then it has no basis whatsoever, which is why the fight by evolutionists is so fiercely against the concept of the global flood described in Genesis in the Bible. If a global flood can ever be proven to have occurred, as stated in the Bible, then the "geological record" as understood by evolution science can no longer be used as its springboard for gauging time. The flood topic is discussed in detail in *The Science Of God Volume 5 - Boats, Floods, and Noah - The Deluge*.

Macro-evolution is major change in form over very long periods of time. Macro-evolution is absolutely and completely dependent upon time and cannot exist without extreme extension of time. Macro-evolution has two key points that it cannot be without, the first is time and the second is the pretended transitions that are imaginatively inserted in to the data and charts and graphs. Just as calibration of radiometric dating is a matter of definition, so too are the found fossils that are typically badly fragmented, causing a need for a great deal of guesswork when trying to place them in cladograms or phylogenetic trees etc.

We of man, who are all Created in God's Image, have an amazing ability to wonder, which without, we would not ever ask anything. When we fail to ask, we cannot retain that which we have never been given the opportunity to contemplate. We as "man" must take full advantage of our amazing ability to wonder and utilize that to advance our understanding and knowledge of all things, especially all things of the Creator. No other known Earthly creature can wonder to a level matching man's ability to wonder.

Chapter 20

What Does the Data Indicate?

Collectively, we have gathered vast amounts of data and fossils. But what exactly does it all represent? Is it macro-evolution, or is it Creation? This book is not about the questions regarding micro-evolution or adaption, rather it is directly addressing the chasm between an accurate view of Creation, versus full-scope evolution or macro-evolution. As mentioned in the last chapter, macro-evolution requires lots of time, millions of years as it is told. Such long ages were *not* unquestionably provable with the geological record at the time this book was written. Many people will claim that the evolutionism time-periods are "provable", but when you do a bit of digging into the details of the subject, you find circular references that generally all point back to Darwin's questionable initial conclusions.

So what hard data do we really have? We do have many fossils, we have observations of existing creatures and existing man, and we have DNA.

Our DNA Data

There are things that we do and don't know about the DNA of man. What we do know is that DNA appears to be the instruction set that causes cells to replicate in a very specific manner that gives them the ability to form into various body parts. What we don't know is how this occurs; we just know that it occurs. While this knowledge is likely to increase in the future, it won't change the reality of how the DNA data in those cells came to be. Was it evolution, or was it Creation? We may never know for sure from a physically testable scientific standpoint.

Since predictability is one means by which we do science, we can make some predictions based upon the Creation model. Here are a few predictions: The brief list of "kinds" stated in Genesis will not be breached by any natural means. Man might be able to pervert those "kinds" at some point, but the "kinds" will always reproduce naturally after their "kind". Another prediction is that man is different than the other creatures and we will see signs of that anywhere man is found. In addition to that, man will have many attributes of God that we would see in Creation and read about in the Bible. And perhaps the most important prediction pertaining to "kinds" and "man" is that, at some point, God had to have somehow infused instruction into living things so that they could reproduce each after their own "kind", and thus we should find instruction somehow embedded into us and all other living things. We do in fact find all of these things to be true and very easily provable.

But when it comes to DNA, many evolution supporters make the case that we share a large portion of DNA with plants, such as grass. If this is accurate, then we will likely see similar levels of shared genes with most other life, be it plant or animal. This tells us that the primary functions of all life must be somewhat common throughout all of life and that those common components are likely largely for cellular functionality and design. The rest of the DNA purpose guides the cells to become the Creation that they are instructed to build. This is clear indication of design as our human experience testifies to. Designers have recognizable techniques much like a fingerprint uniquely identifies you, so too your Creative techniques will in many ways identify you. Is that a reasonable and logical perspective? This only you can decide for yourself. Is DNA a morphed macro-evolutionary product, or was it infused into all life at the point of Creation?

A Scientific Method

When Creation supporters debate with evolution supporters, the evolution supporters often inject "the scientific method" into the conversation stating that something "must be objectively demonstratable, repeatable, observable, and verifiable, using testable facts." This is fair enough of a method, but it is an odd thing that anyone supporting evolution would make such a statement because evolution is none of those, and evolution certainly is not using testable facts.

Evolution does not meet any of these criteria, and any attempt to assert that micro-evolution witnessed in microbiology is proof of macro-evolution is simply dishonest and is an outright lie. Evolution does not use the scientific method. It may use a method and it might be referred to as "scientific", but it is a method largely built on guesswork rather than on actual science.

The Scientific Method

We should all be able to agree that true science will be objectively demonstratable, repeatable, observable, and verifiable, using testable facts. The question then becomes: Is this true of the Bible's Creation account? The objective part will always be something that everyone will struggle with regardless of what side they are stating their case for. But regarding the Creation model, it is fair to say that demonstratable, repeatable, observable, and verifiable are all true without question.

1. Can we demonstrate that plants and creatures will produce seed or offspring after their kind? Yes.

2. Is this process repeatable? Yes, trillions of times per day.

3. Is it observable? If you ever planted a seed and watched it grow or have a pet that had babies or had a child of your own you know without question that it is readily observable.

4. Can this all be verified? It seems suffice to say that no one on Earth would be foolish enough to think that the demonstratable, repeatable, observable aspects do not regularly occur on a daily basis. So, yes, the Bible's account in that regard is verifiable.

5. Are the "facts" in the Bible testable? Well, yes if you consider the Genesis account as factual and apply it to the demonstratable, repeatable, observable, and verifiable aspects.

The Bible easily passes all of these tests, and even the objective part will pass because the people opposed to the idea of Creation cannot deny the demonstratable, repeatable, observable, and verifiable aspects that Genesis so easily passes.

Is Biblical Data Innocent?

The fact that the Creation account of Genesis easily passes the Scientific Method is partly due to its broad nature. Bible detractors can make claims that the Creation account was an afterthought that someone invented to explain things based upon what they see around them. And this could possibly be true, but given the points made in the other volumes of *The Science Of God*, it is unlikely that it was an afterthought. Modern scientific equipment has allowed us to peer into space and confirm many of the fundamentals laid out in Genesis One. But be very cautious here, you simply can **not** reconcile the late post-Reformation

Bibles with real science. The Bible version you choose to use for your analysis will determine the outcome of your conclusions regarding using the scientific method to verify The Bible.

Those corrupted Bible versions mentioned throughout this book have caused the defeat of many good people causing them to turn away from God. Is this innocent? It's hard to say, but it almost looks like an attack from within the Church. But whether innocent errors or not, the errors are errors and have caused untold problems for many people. It is time for everyone to understand that not all Bibles are created equal.

Reading the Data

Your predispositions on the evolution-versus-Creation subject, or any subject for that matter, will ultimately be an enormous factor in your final analysis of any problem that you are trying to solve or any question you are trying to answer. You will read and interpret the data differently if you are for that data, versus if you are against that data.

When Creation supporters have been taught using these perverted late post-Reformation Bibles, they will typically assert that Creation occurred in six twenty-four-hour days. And if evolution supporters evaluate the Bible they will typically not accept or hear any Creation theory that is not a six-twenty-four-hour-day Creation or that does not completely fall in line with their evolution theory. Thus, both sides are often **not** objectively reading the data.

A truly objective person will not have an agenda other than to get to the truth, and they certainly will not use a tainted Bible version from which to get their Creation data. A truly objective person will look at the fossil record, at nature, and at authoritative Bible text and then analyze it all together to see if it all *properly* and logically fits together without forcing any part of it.

Another part of the data that we can read is the logical analysis of our bodies. Just our blood circulatory system alone defies evolutionary possibility on so many levels. You have to study the human body, or even any animal, and consider all of the systems that are dependent upon each other in order for any one of those systems to function. When we listen to evolution proponents talk about these functions, they typically find ways to explain each one of them as a separate unit, but never do they tie them all together as one harmonious functional system. The evolutionary explanations that are offered regarding separate body units are a stretch to begin with, but then when you consider that it is a struggle to combine these organs and systems using even convoluted evolutionary logic, you begin to see how very rare those odds truly are.

The fundamental systems utilized in most creatures have clear utilitarian purpose and are very obviously designed to work *together* as one harmonic system that will sustain and propagate life. Evolution theories break these systems apart and then separately explain, through vague language, how they could have theoretically come about. However, each explanation is a stretch of the evolutionary imagination in its own right. The complexities of hormones, nutrients, blood, various secretions, reproductive organs, circulatory systems, cells, and so much more are all each in their own right far too complex to have come about through chance via natural influence that was without any intelligible guidance. Even something as simple as our fat-energy storage system is far too complex to have come about through natural influence or chance.

If you pay close attention, what you will see the pop-science evolutionist's do is to distract the issues by taking one single aspect of an organ and then find some vague comparison in the natural world that has a very loosely defined similarity to the organ aspect they selected to use for their example. Then they will guide you down a path explaining how the organ might have morphed from the primitive example into the current day organ.

I suppose this deceitful practice's explanation is fair enough of an example for the particular point that they are making in such cases. But each such case is, without question, a stretch. There is no concrete proof of these things in any form as it is all utter speculation. Further, when you consider the odds of what they are suggesting in such examples, you must not take that one example as the overall odds of end-to-end evolution. Odds compound in a very profound manner and in this case those odds compound severely **against** evolution. With each of those far-reaching examples, the odds are so low that they are unlikely to ever have occurred on their own as it is. However, offering the benefit of the doubt to those far reaching suppositions, we still have to combine them all together. So, if something has a-one-in-a-billion chance to occur and another thing has a-one-in-a-billion chance to occur, then the chance of both of those required things occurring together in some sort of harmonic fashion are multiplied in a way that makes it essentially impossible. Now add to this that the body has over seventy complex organs that all work in harmony with many needing each other in order to function. When you multiply the odds for the above mentioned and more to occur purely through macro-evolution, then there are simply too many digits in the number to actually count.

Darwinian end-to-end evolution is solely dependent upon obfuscation and keeping people stupid. If you follow the path of evolution, you are taught to *not* question its validity and that Biblical Creation is all "fairytales". But I propose that it is exactly the opposite, and I encourage people to investigate both evolution and Creation using true logic, truth, and actual data, rather than other people's opinions.

Chapter 21

You Becoming You

A key fundamental of Creation is self-awareness. You cannot exist without becoming self-aware. If you have ever observed a newborn infant during the first few weeks of life, you should understand what becoming aware looks like. While that might seem to imply that babies become self-aware *after* they are born it is not meant to. Awareness starts in man, as it likely did with the **I Am**, at the moment of conception. That is when the child begins his or her **I Am** Image and that **I Am** Image becomes more aware with each new cell and then with each beat of the child's heart. It is a progression of realization that is extremely slow at first and accelerates each day. When the child has finally fully developed and is ready to leave its mother's protective womb, the process of awareness increases exponentially as the fullness of self-awareness proceeds. Within days or weeks they find their feet and hands and begin to take control of their bodies. Their *becoming* begins at the moment of conception. You **became** you the very moment that the spark of life happened as your conception occurred, but your self-awareness, your ability to realize that you **are,** that you **exist**, starts as a slow process that begins with the seed of conception and continues throughout life as the primary concepts are eventually realized by you.

Skeletal Finds

We have the wrong idea about life. As knowledge ebbs and flows, we lose and gain information. You might question the thought of losing information, but society's ability to be dumbed down was on full display early in the twenty-first century. If you look all around the world and see all of the amazing things that ancient cultures built through thousands of years of recorded human history, and then consider the Aborigine cannibals that Darwin discussed, you have to realize that they were not stuck in some isolated area that did not allow them to progress. Rather they lost the collective information of developed society. And according to Darwin, the Aborigines readily accepted the information that the missionaries offered them.

When we find man-like skeletal remains there are a few questions that need to be addressed:

- Are the remains fossilized?
- What were the geological circumstances in which the remains were found?
- What is the approximate percentage of the skeletal remains that was found and recovered?
- What is the proximity of all of the parts of the remains that are claimed to be in the single creature find?
- Are the missing parts realistically modeled to fill in for the actual missing parts?
- Is the skeletal structure categorized properly?
- Has the skull been compared to the vast variation of skull shapes that we witness in living people today?

- What is the angle of degrees the skull is presented at when shown to the public? Is it tilted back or is it set properly upright?

This is the short list of some of the things to question when you are presented with depictions or actual photographs of the various alleged "transitional primates". When it comes to evolution, the story presented is not always the most accurate or most honest. Dig into the pictorials that are offered by pop-science and then ask yourself all of the aforementioned questions in an honest manner. What do you see?

Facts Are Facts

What is a "fact"? That is somewhat dependent upon who you are speaking to. Facts are sometimes confused with truth, and some people reject the concept of Truth and insist that only "facts" matter. In this case, those who reject the concept of truth are essentially saying that only facts matter and only they are worthy to determine what is factual and what is not factual. Truth is a bit different than fact. Truth is what exists, and in truth that is what facts are supposed to be, but somehow that idea has fallen out of favor with many evolutionists.

When you are shown the various depictions of the skulls of alleged primate ancestors, you will notice that they are never accompanied by full skeletons and seldom are even close to full skulls. And in every case, the skulls that are fully in tact tend to be either clearly from apes or clearly from modern man. The in-between fragmented skulls are sketchy at best to begin with. So, the true facts in the case of human evolution are limited. Have we found skulls? Yes. Are some like human skulls? Yes. Do they vary in structure? Yes. Is it fair to imply that these skulls are transitions from primate to man? No, not really.

Actual true facts are going to be obvious once shown and they will not have interpretation attached to them. It will simply be a skull and someone saying "Here, I found this. What do you make of it?" They found a skull and that is an obvious fact because you might be standing right in front of it, and if they are not trying to deceive you, they might have taken pictures of the excavation site showing when and where it was found—All of that would be factual.

Some Facts Are Not Facts

However, some "facts" are not facts. A fact is something that exists, so the found skull is a true fact to be considered. The "fact" that it is a million years old is not a "fact". You will see people get noticeably angry when someone pushes back against their assessment of the not-so-factual "fact" that something is from a million years ago or that it is a "transitional" skull.

Some of the problems with evolution are regarding full analysis of past and current data. The past data that is found in archeological digs is not fully compared to modern skull forms from all around the world today. Pay attention to the various skull forms (their head shape) when you see people from around the world today. There is an extremely wide array of skull shape variation, some of which compare closely to many of the oldest skull fossils found. And if the found skull is not a close match to skulls of modern-day man, then those fossils are always quite primate-like.

The people who are doing the archeological digs are searching only for outliers, so if a typical human skull is found then it is generally disregarded as "modern man", but if an anomaly is found then it is accepted as a transitional fossil. Some of the contention comes with regard to cranial capacity. Some skulls have peculiarly small brain cavities, but as mentioned in earlier chapters, we have to consider abnormalities such as microcephalus where the individual is born with a birth anomaly. This birth anomaly occurs from time to time and those living individuals' skulls reflect the smaller-capacity skulls found that are dated as millions of years old. This is not questioning the age of the skull, but rather the assumption that it is a transition.

The guidelines set forth in the first few books of the Bible are not arbitrarily done. Those were set in place to keep the people heathy and reduce birth defects and disease. God wanted the people to be pure because the "heathen" neighboring peoples had disease issues and God did not want those problems to bleed into "God's people".

When you look into the entirety of the theory of macro-evolution, make sure to pay close attention to the world around you as well. There are many clear and obvious indicators walking around that live and breathe today, testifying as to whether we evolved, or were all specifically Created from the start as "man" in the form that we are today.

Chapter 22

Science and The Bible

Evolution supporters see the Bible as a book of "fairytales", but the Bible is actually quite specifically very scientific. It's understandable that people could have taken the text out of context in the past for two key reasons. The first reason is that they could not look into the heavens years ago or into the microscopic realm as we can do today, so they did not have the advantages that we have in our modern times. The other reason is that man has the disadvantage of being limited to our experiences. So, as we watch the Sun rising and setting it appears as if the Sun is moving when it is actually us who are moving relative to the Sun. So when the Bible talks about a "day", we see that as the only thing we understand a day to be. This is one of the key reasons for the poor translations found in some post-Reformation Bibles that have caused the six-twenty-four-hour-day Creation hoax. The translators of those Bible versions simply did not grasp anything of astrophysics.

The reason this occurs is that we fail to see things through the eyes of others, which is where all issues with our human relationships arise from. Failing to see things from the perspective of others is a very human problem that has caused untold tragedy throughout the world. When it comes to Creation, what good does it do for us to see through to the eyes of other people? Generally, it does little to help at all with regard to Creation, but what will assist us a great deal to see the scientific nature of Genesis One is to try to see things from God's perspective. You can be assured when you take an unprejudiced position where you assume that God exists when evaluating the Genesis One text, that you will arrive at a very different conclusion than if you assume that God does not exist. And then if you read the Genesis text using the Earthly perspective of man, you will also arrive at very different conclusion than if you take the position perspective that the Creator had during Creation. The difference between science and the Bible is that the Bible is history, where part of science, on the other hand, analyzes that Creation history.

It is our choice how honest we each will be in our own evaluation of the Bible's Creation text, especially regarding the Creation of "man". While seeing the Creation perspective through the eyes of our peers does little to help us fully understand Creation, doing so does a great deal to understand why they believe as they do. Life is not always as it seems. All too often, the debates become petty and filled with attacks on personality or credential which has absolutely nothing to do with Creation or with science. So, when you see these attacks you can be reasonably certain that the person doing the attacking is trying to divert attention away from the flaws being pointed out in their own theory.

The difference between being truly objective in analyzing the Creation text, versus typical analysis, is like the difference between light and dark. You are either trying to prove your own theory or you are trying to seek the Truth. If you recall, "science" means to split, which is to make distinction between things to better understand them. Division for the sake of organization is the primary activity during the Creation events. It is in our nature to be "scientific" by making distinctions just like the **I Am** did who we have been Created in the image of.

Who Gets to Define "kinds"?

In evolution science people make distinctions to fit their needs. This is the obvious path of science and it must be done, but who gets to define things? And are they doing it to reveal truths or to set an agenda? The defining of science-words is subjective and the person who discovered the thing being defined generally gets to establish its meaning. Someone who finds a never-before-seen bird fossil can name the species for it or categorize it into an existing species. You don't get any recognition for finding something that has already been found and categorized, so finding a fossil that is different is advantageous to the reputation of the finder of the fossil. Nowhere is this more of a problem than in defining human skulls and defining the alleged "humanoid" skulls.

The battle for definition of both "kind" and "species" cuts to the core of the evolution-versus-Creation debate. Debate is generally good for the scientific community and for society overall, that is until both sides are arguing with false "facts". So, who gets to define "kinds" and "species"? Science should stop using the term "species" because it is not used in a very scientific manner by science. If you were to consider the word "species" as used in the scientific community, you will find that it is not objectively demonstratable, repeatable, observable, and verifiable, using testable facts because it changes with the wind depending upon who is using the word. There are plenty of other words that evolutionary science is using to designate differences in creatures, so "species" may just as well be left out of their debate points.

The term "kind" is utterly rejected by science and is replaced with "species". The term "species" was used in the Latin Bibles many hundreds of years ago, long before modern science came about, therefore by seniority, the Bible gets to define the word "species". And since the Bible is claiming authority by virtue of the telling of Creation before man existed and was able to define anything, the Bible is what will be used. Biblically, the terms "kind" and "species" are equal in meaning and the scope of that meaning is set forth in the Creation text. There are two levels of understanding of what a "kind" or "species" is. The first level is that a "kind" cannot be defined beyond the limited list presented in Genesis One. The second level is that there may be more diversified creatures within any of the Biblically listed "kinds", regardless all creatures will propagate after their own kind, meaning that the offspring will be very close reproductions of the parents. That is how to define "after their kind". The offspring will always be an extremely close replication of the combined parents. This is true with animals, but it is even more stable with "man".

Believers versus Christians

What are the rules or parameters to be a Christian? Who sets those rules as to what a Christian is? I understand that when you attend a particular church, you will get a basic set of understanding, however there are many different Christian belief sets and some of those have inaccurate perspectives. But, that does **not** mean that **all** "Christians" believe the same information. The book *Understanding The Church - Upon This Rock I Will Build My Church* describes this problem very well.

Much the way some scientists believe the Universe is expanding, and others believe it is contracting, some Christians will believe the Bible says one thing, and some Christians believe the Bible says another regarding Creation. For instance, some people believe the Bible implies six twenty-four-hour Earth days while other Christians believe those days are actually long epochs or extended events.

Christianity has little to do with Creation, but it is the Christians that are largely responsible for carrying the Bible through the past couple of thousand years. Jews have similar beliefs regarding Creation, but most Jews don't accept Jesus The Christ as The Savior. So both are believers, but only those who believe that Jesus The Christ was the Savior are considered to be "Christians".

Evolution is filled with believers of evolution. Believers are everywhere. If you did not believe that there was going to be a floor under your feet when getting out of bed, then you likely would not attempt to get out of bed. You believe that the floor exists, therefore you proceed with your experiment of placing your feet on the floor. When your feet make contact with the solid surface then you know that it is there and you will walk away from your bed to start your day. But imagine being in your bedroom at night and all lights are out and you can see nothing. You left the door open on your way in, so you get up to do something and, as you leave the room, **Bam!**, you crash face-first into the door that your spouse or kids for some reason shut without you realizing it. In this case you *believed* the door was wide open, but when you tested that in the dark, you found that the door was not open.

Science is no less of a belief system than is religion, and based upon the function of blind faith, evolutionary science is more of a blind "religion" than is Christianity. At least with Christianity and the Creation account in the Bible, there are documents that are widely corroborated and they also agree with nature and archeology. The same is not true of evolution science. There are many people in the evolution science field who believe that their evolution door is wide open—but it is not. When they run into that closed door they force it open and then they go on to pretend that is was never closed. Inventing theories does not make those theories true just because they are claimed to be "science" or "factual".

Is the Bible Scientific?

We touched on this before. The Bible's Creation account is provably objectively demonstratable, repeatable, observable, and verifiable, using testable facts. Let's test this:

Do the limited "kinds" listed in the Bible exist? Yes

Do we see those "kinds" everywhere we look around the world? Yes

Can we verify that this is so? Yes

Can this be tested? Yes, through the process of verification by actually looking at nature and society.

The same is true of man.
Does man exist? Yes

Is man unique? Yes

Does man have dominion over the animals? Yes

Does man have a companion like him that is a woman? Yes

Were they fruitful? Yes

Did they multiply and fill the earth? Yes

You can do this process with any part of the Genesis Creation text when you use an authoritative Bible version and the Bible's Genesis account will pass the so-called "scientific method" every time.

There is nothing in the Genesis Creation account that cannot be verified by simple observation. The opposite is true of ape-to-man macro-evolution.

The Spirit of God

The Creator is the ultimate scientist. And since God the Creator is Pure Thought and Pure Spirit, true objective science is a spiritual endeavor to better know the Spirit of the Creator. If you have ever watched one of those frustrating, and generally quite annoying, movies or television shows where a person is lying and they have to keep adding to their lies in order to keep up their ruse as they plug each hole in the previous lie, then you can understand pop-science's evolution. That is exactly what is done with the evolution theory. It is a lie built upon a lie built upon a lie that just keeps getting the theory in deeper and deeper trouble, and eventually that house of cards is going to come tumbling down with a very violent crash, crushing and utterly devastating its occupant pop-scientists.

Man's unexplainable evolutionary superiority over the animals perplexes far too many people. But interestingly on the Biblical side, the answer is really quite simple, we are Created in the **I Am** Image of the Creator and we have many of the same attributes as the Creator. This is not true of the animals or any of the "primates".

When you buy into the notion that there is no Creator, then you must invent scores and scores of lies in order to even attempt to explain our origins. But when you accept the Creator as stated in the Bible then all things pull together and work in harmony regarding our origins and the origins of all of tangible existence. And it all agrees with nature, physics, logic, reality, and true science.

Chapter 23

Ask and You Shall Receive

In an earlier chapter, curiosity was discussed. Curiosity is one of our most important gifts, but we commonly run into walls with our curiosity because we fail to properly form our question, ultimately failing to understand what exactly our quest is. Jesus The Christ said "Ask, and it shall be given you: seek, and you shall find: knock, and it shall be opened to you." Whether you believe that Christ existed or not, and whether or not you accept Christ as your Savior, will not take away from the fundamental truth in the statement "Ask, and it shall be given you: seek, and you shall find: knock, and it shall be opened to you." In the Old Testament it says "My people perish for lack of knowledge." Those two statements are quite profound.

If we fail to properly form a question and then ask it, and fail to further seek the answer by actually going out to investigate by knocking on the door of understanding, then we simply will not learn. Someone could assume that if someone tells you everything, that then you will know everything, but this is not true in any respect. We generally only retain that which holds our interest. Whether it be the Biblical or evolution sciences, when you read the documents, ninety-five percent of the information flows right on past you, even though you read every single word of it. But the questions that you formed in your mind beforehand will cause the answers that you read to stand out in the text in a way that immediately catches your attention, thus making them memorable for you. So, if you are an evolutionist who discounts the Bible as "fairytales" or a six-day Creationist who refuses to accept any form of "evolution" or old earth Creation, then you will be blinded by your stiff-necked ignorance and will be unable to form the proper questions that are needed for you to ascertain actual truth.

Every major discovery has been made by those who formed questions in their mind and who then asked those questions and went to seek the answers. They knocked on the door by digging for fossils or by testing aspects of physics to prove their theories. Or they looked to the heavens with a telescope and found new truths that others did not ever even think to ask.

Our ability to wonder is exercised only when we "ask, and it shall be given you: seek, and you shall find: knock, and it shall be opened to you." Perhaps the single most scientific act "man" can do is to "ask"

Expectations

When we properly form a question in our mind, we immediately have some expectation for that question, and it is those expectations that we typically seek. If our expectations are not realistic or are not rational, then it is unlikely that we will find what we expect. But if our expectations are realistic, then often we find more than we asked for but only if we make the effort to seek. For instance, we safely say that there are billions of galaxies in the Universe, but while that is likely true, it is also likely extremely shortsighted. The estimated visible Universe is only what we have thus far been able to see due to our seeking attempts. But because we actually looked for *any* galaxies we ended up finding more of them than we ever imagined. Yet, given what man has witnessed in the heavens, it is far more likely that there is no end to the galaxies. Space is almost certain to be infinite as far as man will ever be able to see, and the amount of galaxies is likely infinite as well. Whether there is a Creator or not, we all will eventually come to conclude that "matter" in the universe is endless and it is infinite.

When it comes to the arrival of man and the topic of evolution, we have the same basic thought functions as with when the heavens were Created–we first have to ask. But when we ask the wrong questions or ask questions for the wrong reasons, we get wrong answers because the wrong answers are what we insist upon finding, and that insistence blinds us from the right answers, all of which allows us to believe our own lies and inaccuracies. In studying Genesis, far too many Creationists and evolutionists read the text with similar prejudice as one another where they know that a current era day is twenty-four hours and thus to them it was all made inside of a week's time according to their incorrect Creation account interpretation. Six-day Creationists accept this, but evolutionists rightfully reject the utter ridiculousness of the six-day idea. Yet, both sides fail to ask the more obvious question regarding the order of events listed in Genesis: How do you explain "heaven" and "earth" having been Created before "Heaven" and "Earth" were Created?

Our interpretation of the order of the Creation events forces us into one of three camps: "The Bible told me so!", "It's all bullshit!", and finally the more logical, which is, "what are we not understanding properly?" If someone does not believe in God then nothing really matters and the Bible is just stories. But if you truly believe in the Creator that you claim to believe in, then it's time to stop believing in an *illogical hocus-pocus* type of God who instantly Created all things seen and not seen–which is a perspective that is nothing but irrational.

If we are going to take the position that a Creator Created all things, including "man", then we are forced to realize that the

Creator is highly **rational**. And if we derive our beliefs of God from the Bible and believe the Bible to be the "inerrant word of God" then how can we believe a Creation account that does not perfectly align with that very same Creator's known physics and nature that we all live in that **I Am** Creator God Created?

We simply cannot unless we are being irrational. If we accept the six-twenty-four-hour-day Creation belief then we are certain to need to invent sub-theories in order to justify our lies just like in those aforementioned movies where people keep building lie upon lie. And the same is true of evolution. If you accept macro-evolution as realistic, given what we see all around us, and if you completely discount the Bible as "fairytales", then you will be forced to invent things to justify your lies. We see this all the time from both sides of the debate. We should have a great deal of admiration for those who are unsure and stay tight-lipped and are waiting for truth to appear to them. For that is far better than leading people down an erred path of deceit.

Both the six-day theory and the God-did-it-through-macro-evolution theory are contrary to God and contrary the very common sense instilled into us at our conception. God makes stuff—God doesn't make stuff up. The Bible is either false or it is not, and if it is not then it must align with all that we see in nature, with no hocus-pocus POOF! allowed.

In evolution, we have this view that because we share attributes with other creatures, that then "proves" that we are related through descendancy. But living structures, including our bodies, are nothing more than extremely sophisticated machines, and it is the breath in the creature and in man that makes us more than mere machines. We can look at automobiles from around the world and assume that the all cars evolved, but we know better. The reason that cars share many attributes, such as tires and engines and transmissions and bodies etc., is because those are basic well-designed requirements and are parts that work and are versatile. The same is true of creatures. We share attributes because those attributes work for the particular functions that they are designed for. If a God did Create all living things, then based upon every part of our real and tangible life experiences, we would expect that God would design systems to be common in the creatures. And in fact, from the DNA all the way to the finished product, that is *only* what we see. We see pattern and repetition, there are no arbitrary occurrences. And nearly all birth defect deviations occur due to environmental circumstances.

All organs in a body, whether of animal or of man, are of common design and perform specific tasks and functions to keep the creature alive. This screams of design everywhere you look.

Earliest Primates

The earliest primates are said to have evolved roughly sixty-million years ago, about five million years after the demise of the dinosaurs. If you grew up in the city it is understandable that you might not have firsthand knowledge of how erosion works, but if you grew up in the country, especially if you grew up on a crop farm, you should be very familiar with how fast erosion occurs and how quickly rock can break down in weather, especially during freeze-thaw cycles. Sometimes erosion takes a while, but if soil and rock changes at all during *your* very short comparative lifetime, then how likely is it that a span of five million years between dinosaurs and the first primate would leave any evidence at all, especially relatively shallow footprints in rock that we find on or very near the surface? A dinosaur footprint that is allegedly sixty-five million years old should no longer exist unless it has been buried until the last several hundred years at most. We should all be able to accept that. And also, how then do we explain human foot prints being in the same strata that is believed to have come no less than five million years after dinosaurs, according to the macro-evolution theory timeline?

Something is not right with these timelines. There are cave drawings and ancient figures of dinosaur-like creatures; there are the statements in the Bible in the book of "Job" that give details of creatures that would easily be cast as a dragon or a dinosaur. We have medieval tales of dragon slayers. We even have instances of alleged dinosaur and human foot prints together in the same rock piece. Either there has been a multi-thousand-year conspiracy to deceive the world, or this common thread is trying to tell us something. The earliest primates may in fact have been about sixty-million years ago. But they are not related to man in any genetic way whatsoever, plus man's appearance on Earth is very recent in relative terms, and sudden according actual geology and science.

What we do know about man is that we have recorded lineage back to Adam and Eve after they were ejected from the Garden of Eden. And that lineage spans roughly only six-thousand years. This timeframe is consistent with archeological digs and nearly all surface-exposed skeletons that have been found, except of course for those that pop-science has chosen as its "proof". Evolutionists are going to have to do better to convince rational people that macro-evolution is real. Misdated skulls and badly fragmented skulls and clearly mis-categorized skulls are not evidence, they are guesses and/or agenda. Similarly, the six-twenty-four-hour-day Creation side of the debate is going to have to do better to get rational people convinced that six-twenty-four-hour-day Creation is real. "God says so" and "God made it look old" are not rational thoughts given the evidence that surrounds us all and given the as-a-matter-of-fact nature of God as told in the Bible.

It is possible that the earliest primates came about sixty-million years ago, yet humans have a very recent arrival based upon the Bible, as well as based upon what the true archeological evidence shows. Regarding archeology and time estimates, how can we trust an industry that has an error margin of plus or minus five-million years? The current estimates for the first primates run between sixty-five and fifty-five million years ago. That is a ten-million year spread. That is a margin of eighteen percent on the high side and fifteen percent on the low side. If the evolutionists have a plus or minus five-million-year margin, ten-million years total at the sixty-five and fifty-five-million-year age range, then what is to say that they don't have a similar ten-million-year margin on more recent life? This means that if they say something was ten-million years ago it could easily be only several thousand when using their age estimate tolerance.

It seems a waste of time to even address the ridiculous proposals that come from macro-evolutionists or from six-twenty-four-hour-day Creationists. But since these two perspectives have both been equally damaging to the world, this needs to be immediately addressed head-on in order to stop their lies that have resulted in the hemorrhaging of Christ's True Church.

Believe what You Will

We can choose to believe whatever we want to believe, and we can accept any evidence that suits our beliefs, but in the end, we all still have to deal with Truth. Eventually the Truth will be revealed, and when it is, you must ask yourself: Do I want to be defying Truth and not abiding by that Truth when the Truth is revealed?

When things are finally proven, it seems fair to say that both the macro-evolution folks and the six-twenty-four-hour-day Creation folks are going to feel pretty foolish for the lies and misinformation that they have been feeding to the general public. Only things that are true are true, and no matter how badly a person might want to be right in their current beliefs, nothing we do will make our erred beliefs correct other than to actually be correct. You cannot be correct if you fail to seek Truth about the Creation of man and about evolution.

Chapter 24

Final Analysis

In the final analysis of the evolution-versus-Creation debate, we find that there are two levels on each side. Evolution has *macro* and *micro* and Creation has *six-twenty-four-four-day* and *long-age*. Long age Creation is often improperly used in an attempt to make the Bible reconcile with macro-evolution so that they can hold onto their God **and** their science. But long-age Creation has absolutely nothing to do with evolution. Could it? Possibly, but it is unlikely. For all of the reasons mentioned in this book, macro-evolution is obviously false. But what about six-twenty-four-four-day Creation, is it a possibility? No, it is not a possibility. If we choose to believe in the idea of Creation, then we also have to believe in a Creator who Created it all. And to make the assumption that a Creator who so brilliantly and proudly Created "light" is going to cheat the Creation of light itself to get light to us sooner than it could on its own from so far away by defying the Creation itself, is utterly illogical! And to further believe that the Creator would POOF! creatures into existence defies everything we see in nature and thus is also utterly illogical.

Man has been given a very unique gift of logic and reason that is so far in excess of any animal that it is a sin to not utilize that gift. In Jesus' talents story in Matthew 25:14, the talents are referring to money where one of the men buried his talent and when the master returned, the master called the servant a "wicked and lazy servant" for not increasing his talent and instead hiding it in the ground. Truth is an eternal promise that we must "ask, and it shall be given you: seek, and you shall find: knock, and it shall be opened to you." Here I would like you to think of talents as your actual talent of the ability to use reason and logic. We all need to seek Truth, rather than our own agenda. And, we all need to increase our talents through truth, rather than burying our talents in lies.

Intelligence of the Ancients

If we actually evolved and are now in a higher state of evolution, then why do we look back at all? Should we not rather look forward to see where we can go? Ironically, when we look back we find that many of the ancient people had very deep knowledge of the detailed workings of the Universe from very early on, and in modern times have only recently re-discovered many of those same things. Why is this?

Darwin's natives that were allegedly cannibals, and very much not culturally sophisticated, were considerably less advanced than people written about in the Bible several thousands of years *before* Darwin. How did man lose this information and become so savage? This speaks volumes of the accuracy of the Bible and the ancient extra-Biblical writings. Since the Bible states that man was made in the Image of the Creator with abilities to Create, we then can make some assumptions based upon that information.

We can assume that some fairly organized quite intelligent ancient pre-flood cultures would have existed. They might not have had things like microchips figured out or any other technological ability to communicate as we do today, but they would have flourished with abilities to build cities and thrive in them until they had transgressed against their Creator. This means that at some point buried deep beneath the layers we will find such city ruins and when that occurs, it will confirm the Bible's flood account and finally put to rest the overreach of macro-evolution. The evolutionary thought that humans were stupid in prehistoric times is absurd based upon archeological finds that we perform in our modern times, as well as according to the Bible. Humans have no "prehistoric" times. "Pre-*historic*" means before history, but before the history of what or who? "Pre-*historic*" means before *human* history, because only man has been proven to record history by actually writing it down.

Based upon the Bible, we can make predictions, such as someday we will find ruins of pre-flood communities buried in layers of debris and sediment deep beneath the surface of the ground. We will never find any definitive missing links that clearly prove a transition from primates to man, nor will we ever find clearly proven transitions from any Biblically stated kind to another Biblically stated kind. We will find no deviation from the current form of man, nor will we find parallel evolution of man such as what we see in animals like with the vast array of birds. Man will be shown to have originated from one area and will have spread abroad over time. Man will be very prolific and quickly populate the Earth.

Dangers of "science"

Science is good, right? That all depends upon how we define science and how we use science and how honest we are about it all. If there is a God then there is no question that science is a God-given quest which we are to pursue since we are made in the Image of the Creator. We are told to "ask, and it shall be given you: seek, and you shall find: knock, and it shall be opened to you." When we do "science" we should be following the scientific process as set forth by God. That is to say to, *investigate, consider, plan,* and then *Create* for the increase of all good things.

Science becomes a dangerous tool when we use it to destroy or to deceive. Take for instance all of the weapons of war, these are all Creations meant to destroy and cause otherwise good people to destroy each other. Or consider the lies of science and how damaging they are. First, consider the science of interpretation or translation and how that can be so terribly perverted as is discussed in *Understanding The Bible - The Bible How-To Manual* AND *The Things We Don't See*, or how those corrupted Bible versions can deceive us as revealed in *The Sc The Science Of God Volume 1 - The First Four Days*. And then further how poor interpretation or translation brought about the deceptive theory of macro-evolution, which brings about atheism. Now if there is no Creator, then God simply does not exist, and at that point who cares about all of that, right? Well if God does exist and the Bible is true, then man has a serious problem to contend with. Souls of man all have the same indestructible ever-aware type of conscious and will spend their after-death eternity in a very tormented unhappy state-of-being when those souls make wrong choices.

Everything we think, every decision we make, and every action we take, are all forms of science. Just because what we are doing isn't something that "requires" a college degree, does not mean it is not scientific. The so-called "scientific method" spoken of in previous chapters is not science itself; it is only one way of testing science.

Wonders of Science

True science is done by everyone. Some of that science is more notable than other science. When we look to the heavens and see the wonders of The Science of God there is no end to the discoveries that we can make. But with man's God-like "in the image of" Created nature, we should continue to look within to study the wonders of God's Science within us. We can actually see the programming that God infused into us when the Creator sequenced our DNA. And even when looking outside of ourselves there is an abundance of data all around us just shouting and screaming for us to pay attention to it all, and all of it testifies to a deliberately Created existence.

Scientific Salvation

We always look at Salvation as this sort of obscurely reasoned gift from God that we must accept or we will perish in eternal hellfire. And while that appears to be a part of the general message in the Bible, we must ask, "Is there a somewhat scientific function behind it, or is Salvation only because God arbitrarily said so?" Science gets a bit less tangible when we stop speaking of the physically and scientifically detectable aspects of the physical realm. However, since a soul is *something* and is brought to awareness through being physically conceived, then is that soul not a grand scientific achievement? It is indeed a wonder of science, just as God making man in God's Image and breathing into man the breath of life was a grand scientific achievement. That particular scientific experiment has been successfully performed many billions of times throughout human history.

Now, if those amazing souls of times past still all exist in a non-physical realm, but can still think, are they any less scientific because the souls might no longer inhabit a physical body? Obviously not. Science is not some physical procedure, it is a soul or spiritual activity that our soul/mind/spirit does through our body in the physical realm, and so all of the logical processing actually occurs in the *intangible* realm.

When our Father Adam and our Mother Eve defied God, something scientific happened to them, but it wasn't good. There is no indication that they ever wondered about the Tree of the Knowledge of Good and Evil, that is until the serpent came to Eve and caused her to touch the fruit that they were commanded not to touch. According to Genesis, this caused some sort of change in them which caused them to be less than they were

Created as. Somehow this corrupted them so that they were unable to remain in paradise.

God promised them to send his Word, that's the same Word that is the **I Am** which Created everything. God said that the Word would shed his blood for them. In the Bible, we are told that "the life is in the blood." This is consistent with science. God breathed the breath of life into Adamah. Our blood carries our breath throughout our body. When we breathe, our blood becomes cleaner. Take away the breath or the blood and we quickly die. That confirms that the life is in the blood. So, since we are man and our breath circulates through our entire body in a very calculated manner, it is fair to say that life is also a scientific process—Life is a pattern that has stood the test of time. Since the Bible makes such a big deal about the blood, we need to take it seriously if we believe in God.

We tend to disregard science the moment we talk about the intangible spiritual realm where the soul resides, but since science is a mental process more than it is a physical process, we have to take seriously the reason Christ Bled on the Cross. *Understanding The Church - Upon This Rock I Will Build My Church* discusses how God promised to take on the things that happened to Adam at the time when the Word would come to save Adam and Eve and their righteous offspring from there transgression in paradise. But here we will get a bit more into the intangible aspect. Was the blood merely symbolic? It certainly was a sign or symbol, there's little question about that. But was it done for technical scientific purposes, or was it just some random arbitrary decision that God made? To try to answer that question we have to go back to the intangible fundamentals of the **I Am**.

While we will probably never fully understand the **I Am**, we can get a general overall idea. The overall **I Am** Creator has no physically tangible aspect, and all of the **I Am** Creator is pure scientific thought. Our Salvation has to do with being *at one* or *atoned* with the **I Am** Creator. So our Salvation is not a physical thing the way we tend to see it. Salvation is a scientifically mental process of thought with our acceptance of the Word of the **I Am** in the Blood of Christ. Accepting Christ's Blood is to accept the Word, which is the **I Am**. We accept the physical Blood intangibly with our thoughts, but it is really the life in that blood that we are choosing to follow. Genesis says "Therefore a man shall leave his father and his mother and hold fast to his wife, and they shall become one flesh." When they Created a soul at the moment of conception, that unique soul is a unique combination of the souls of the mother and of the father and the new soul resides in cells that are being built through the mechanisms of Creation. **We** are that soul who is Created by our parents, and we all share the breath of God through our parents that was first breathed into Adamah.

When Adam and Eve touched the tree, they scientifically tore themselves, that is to say their souls, away from the **I Am** part of them, away from God, and placed it into Satan's care. Satan has always hated man because of Satan's jealousy of Adam. So when Adam and Even submitted themselves to Satan, they become one with Satan rather than being at one or **atoned** with God. They made an intangible choice to tear away from God and attach to Satan.

Since this is not a physical process, it is a bit more difficult to grasp because we are only familiar with the tangible, so generally our experiences relate only to the tangible. But if you can set aside the tangible and think about *thinking* or *thought* and how your thoughts are attached to other thoughts then you can begin to get a grasp on your **I Am** Image-soul being at one with The Great **I Am** Creator God.

To put this in a more physical depiction, picture yourself on a ship in stormy seas. The captain says don't go by the rail or you will get washed overboard in the storm, and you say "Okay captain." But then along comes a merchant who says "Why don't you go over there and check it out?" But you say "The captain told me not to." So the merchant says "Nothing's going to happen." Then you say, "Okay, cool!" and you disobey the Captain because you're curious, when suddenly a massive wave washes you overboard in the tumultuous storm. Now you regret disobeying the Captain and realize that he was trying to protect you. In your massive embarrassment you want to drown in the waves as you bob up and down. The Captain then realizing that you disobeyed his orders shouts out for you, and you yell back "here I Am, over here!" Then the captain grabs the life preserver and a rope and tosses it out to you. But you refuse and say "I don't believe in that shit!", so instead of being saved you drown and sink into the abyss, yet you don't exactly die, instead you are forevermore conscious of your foolish rejection of the captain's lifeline, and there is no way for you to return.

That life preserver is Jesus The Christ and the Rope is the Word of God residing in the Blood. If you reject the Life Preserver and The Rope that can pull you into safety so that you are once again with the Captain enjoying yourself on the ship, you will perish. Your **I Am** Image existence that began at one with the Great **I Am** Creator is an intangible scientific matter of choice that only you can make. If you fail to grab the Life Preserver and be rejoined to The **I Am** Creator, you will forever be bound to the abyss of Satan in hell. Salvation is your scientific choice and privilege.

I do not believe that there is a single soul that has ever existed that when told about God didn't believe to some extent. But there are many who reject God by professing that they do not believe in God. When someone takes this position, they are barring themselves from grabbing the Life Preserver. Their arrogance causes them to turn away from the Life Preserver and say "I don't see any life preservers." when they know full well that the life preserver is floating right next to them and is easily within reach.

But there are other people who got washed overboard on the other side of the boat who are now drowning and no one seems to have noticed. Since no one is helping them they might drown and be sucked into the abyss. If you choose to grab the Life Preserver, you are then also responsible to take that Life

Preserver and throw it out to other people who no one is paying attention to so that they too can be pulled into safety. So you decide to grab the life preserver, and then as you climb aboard the boat, the merchant immediately beings to distract you with all sorts of sales pitches and news. This causes you to forget to throw the Life Preserver out to others who are also in great peril. Salvation is intangibly scientific and it is the process to once again be joined with the Great **I Am** Creator which can only be done through tightly grabbing hold the Life Preserver of Christ *and* Rope of the Word in his Blood.

We Are Significantly Unique and Special

You will hear a lot of catchphrases when you follow the evolution-versus-Creation debate, such as statements like you are "*in*significantly unique and special in the Universe." I cannot recall who said that little poetic phrase, but it is wrong. Man is very *significantly* unique and special in the Universe. Even though the likelihood that there is other identical "man" life than just us "made in the image of" the Great **I Am** Creator throughout the Universe is about one-hundred percent, that fact does not take away from our *significantly* unique and special nature.

If you have ever paid attention to culture, it is when we turn from God that we lose our uniqueness. Often we follow wicked ways, and in so doing we copy those who we see as "cool", thus becoming cheap imitations of them. We look like them, we act like them, and we talk like them because we are taught that that's unique and that they are "the cool gang". They might be different from the general populace, but they are not unique, and being like them defies your unique God-like nature, thus causing the person following them to be more Satan-like, which can be immediately erased by grabbing the Life Preserver and Rope at any time.

We see this copying to be "cool" in many aspects of life, but one prominent area is in the colleges with regard to Creation.

Similarly on the six-twenty-four-hour-day Creation side, we are only repeating things other people say. It's okay to say something that someone else said when what you say is your own thoughts about it. But when we just learn the words without understanding those words, then we are nothing more than a cheap imitation parrot just copying everyone else.

"Man" clearly did not evolve, and we are, without question, separate and apart from the animals just as is indicated in the Bible's Genesis One Creation text. We are all Significantly Unique and Special when we grab the Life Preserver and The Rope. But all too often we grab the Life Preserver but we let go of The Rope, leaving ourself to float adrift just above the abyss, until someone hopefully finds us and rescues us again with the Life Preserver **and Rope**. You might be unique if you grab the Life Preserver, but without that Rope you are only a cheap copy of society and a false Christian—and you will perish.

You are Created Man who is a Spirit Being Made
In The Image of God Who is The Great I Am!

Announcements

Theoretical Physics for Everyone!

Can we go back in time? Can we break the light-speed barrier? Is our Sun turning into a black hole?

Explore the secrets of the Universe with a new approach to science that sheds a greater light on pop-science. Mysteries of the Universe are revealed in this easy to understand insightful, in-depth, and thought provoking book about the science of astrophysics.

Theoretical physics can be more than mere theory when the theory is sound. You don't need to be a rocket scientist to understand most of physics; everyone is welcomed in the quest to discover the mysteries of the Universe!

This breakthrough book exposes errors of modern science in the same way that Copernicus, Galileo, and Newton did centuries ago. Are Einstein and Hubble amongst the group of gifted minds that set forth our understanding of the Universe, like those of centuries past. Or are Einstein's and Hubble's theories wrong? Explore these and other questions in *Bending The Ruler - Time Travel, The Speed of Light, and Gravity* and learn how to become one of the great minds that discover the mysteries of the Universe!

Search: Bending The Ruler Book
SayItBooks.com

Announcements

Take Back Control of Your Life.

If you feel stuck while life unfairly drags you down, then now is the time to take command of your life and learn how to overcome the source of your troubles.

Those around us are often those who hold us back from living rich and robust lives. Realizing that those around us are often those who hold us back helps us to understand somewhat, but in order to free ourselves from their grasp and break the chains that bind us, we need to know *why* this happens.

Cut to the core of your problems with *Hot Water* as it walks hand-in-hand beside you through each detail of the cause of problems while exposing the dirt that society buries us with. This thought provoking book explains the details and how most troubles come to be so that you can better understand what to do about it, allowing you to take the control of your life away from those around you to place it firmly back into your own hands where it belongs.

Advance to your next place in life and richly and robustly live your life filled with wealth and joy. *Hot Water And Your Perceived Identity* assists in gaining full control of your life to change your future forever!

Search: Hot Water Book
SayItBooks.com

Announcements

MARRIAGE MANUAL
MAKE YOURS A
Red Hot Marriage
Made In Heaven Filled With Passion and Joy

Learn the Secrets to a Successful Marriage

Have you been trying unsuccessfully for years to tell your spouse the way you truly feel? Are you suffering in a lackluster marriage? Is your marriage on the rocks? Are you planning on getting married in the future? If you answered yes to any of these questions then *Red Hot Marriage* is for you! This straightforward book covers these and many other common marriage problems and also reveals the causes and solutions for some problems that are not-so-common.

The information in this powerful book, like a true friend, can be at your side with each step you take in restoring your life and relationship to where you likely imagined them to be.

We all deserve lives filled with joy and passion, but our relationships have been tainted by society and by our upbringing. *Red Hot Marriage* strips away all of the lies that we have been inadvertently taught, and quickly teaches you how to regain control of your marriage so that it can be as robust, fulfilling, and passionate as you expected. The mysteries unveiled in *Red Hot Marriage* can have you in command of your marriage in short order as friends and family watch in amazement while you and your spouse walk the path to a strong, vibrant, healthy *Red Hot Marriage*!

Search: Red Hot Marriage Book
SayItBooks.com

Announcements

THE FAMILY MANUAL
HOW TO BUILD A
STRONG FAMILY
A FOUNDATION OF ROCK

Building A Strong Family Is Easy When You Know!

The world has us believing that, somehow, we are now different than people were as little as fifty years ago. With all of the emphasis on modern behavioral disorders and the mass misdiagnosis of pop-culture diseases, parents have few places to go for information that is true, insightful, and trustworthy.

Strong Family explains, in detail, how family life slowly becomes tainted to a point where our children too often become rebellious and, sometimes, even unmanageable. This even happens to parents who are very loving people.

How to prevent these issues from occurring in the first place is explained in *Strong Family*. But more importantly, *Strong Family* explains the details about how to stop it from progressing further and even how to reverse the damage. *Strong Family* takes a no-nonsense approach to revealing the secrets and mysteries of how parents raise smart, productive, healthy children.

We all deserve joy and love in our family life. Intelligent, healthy, kind children are a right that all parents have, but without understanding the details explained in *Strong Family*, the quality of your children is left to chance and your rights are forfeited. Don't roll the dice with your family. If you want to know the secrets to unlock the mysteries and solutions to a great and joyful family, then *Strong Family* is for you!

Search: Strong Family Book
SayItBooks.com

Announcements

The Prayer How-To Manual
Understanding Prayer
Why Our Prayers Don't Work

Learn the Real Secret of Prayer

There's a secret that many have tried to understand but failed to accomplish. We pray day after day after day with little or no positive results, causing us to lose faith.

Some people believe that there's a secret method that must be followed to get your prayers answered and receive the things you want in life, but their success is limited, if it comes at all; while others believe that they're not worthy to have their prayers answered. Few people know the True secret, and when they tell us we often misunderstand them.

Understanding Prayer explains, in easy to grasp language, the mysteries behind many causes of prayer failure. True success in your prayers is not measured by how often you pray, how long you pray, or even how badly you want something and how hard you for pray it. True success in your prayer life is measured by *results*!

Understanding Prayer offers you the opportunity to get those results as it reveals the mysteries of a full and robust prayerful connection allowing you solid and repeatable results nearly on command. A little time to read and pray is all it takes to quickly put these sound, true, simple principles to work for you and your family. Gain the understanding of prayer and of how to receive the blessings of financial and mental wealth that can benefit you and keep you free from strife and trouble for years to come!

Search: Understanding Prayer Book
SayItBooks.com

Announcements

When You Dream... DREAM THIN™
The Weightloss Repair Manual

Learn How to Lose Weight While Sleeping

How many people do you know who exercise and still can't seem to lose weight? Has that ever happened to you? As a matter of fact, because we don't know the vital secrets that are shared in *Dream Thin*, many of us actually end up *gaining* weight when we exercise.

Do you hit your weight loss goals? And does your weight stay off when you do actually lose some weight? Even many doctors miss the *real* answers to weight loss. If you doubt this, then simply look at the waistlines of many medical doctors and nurses.

Weight loss is easily mastered when you understand a few basic principles. We often go on fad diets or follow the orders of our doctors, only to put the weight back on even faster than we lost it. Many of us suffer from unnecessary disease, and some of us will die too young.

Dream Thin does more than simply share answers to weight loss mysteries. *Dream Thin* explains the important details of *why* and *how* weight loss connects to mind and body. The information in *Dream Thin* allows you to make weight loss permanent without having to try so hard. Don't make more of the same empty promises to yourself each New Year's Day. Instead, quickly and easily change things today and make all of your tomorrows better with *Dream Thin* while still enjoying all of the foods you eat today—and yes, even fast foods!

Only you can choose if you want spend your hard-earned money on medical bills and funerals, or if you would rather spend your time and money looking great while being out and about and enjoying life with friends and family as intended!

Search: Dream Thin Book
SayItBooks.com

Announcements

How to Win When You Think You've Been Beat

Feeling grateful is a bit of a struggle when we face tough times in our own lives, and avoiding depression during those times can be tricky. The world cares little of us when we face our own personal struggles, in fact life kicks us when we're down. You've probably experienced the world caring little of your past or present problems, so looking to "The World" for rest and peace is typically of little help.

It doesn't have to be this way! You can change your disposition, and thus, change your future! It's no big secret and it's not difficult, but "The World" won't tell you that, so very few people ever get to hear or understand this simple "secret".

It's amazing to see the people and situations you can attract into your life when you find your own proper perspective, and once you find it you will not want to let it go! Days that test you to your limit become far easier to overcome, making every future test easier than it otherwise would have been.

Simply understanding a few key basics can change your direction in life in short order and can make life a whole lot more peaceful and Joyful! Let *Thank You God – Finding Gratitude in Hard Times* be one of your keys to peace and Joy!

Search: Thank You God Book
SayItBooks.com

Announcements

Volume 1 - The First Four Days

Is there a God? Did we evolve? Did everything start from a big bang? These questions have been plaguing our minds for many years. Only science-minded people and clergy seem to have the answers. But do they really have any true answers?

Is what we are told by science true? Is what we are told by the Church true? Or are there other better explanations for everything? Did we hitch a ride from Mars, or is that all fantasy science? Was everything Created in six twenty-four hour days, or did it all take billions of years to happen? Few people are willing to even fully consider these questions, and even fewer have any coherent answers. *The Science of God Volume 1 — The First Four Days* challenges your current beliefs while asking tough questions of science and of the Church.

For years, Christian after Christian has attempted to argue for God and the Bible's Creation only to fail miserably. Why is this, why is it that Christians cannot seem to win this debate? Often Christians think they are winning the debate only to find themselves at a loss to answer the real questions, and then they get mocked for their poor answers.

Whether you are a scientist or an average Christian and want to discuss the Creation debate, *The Science of God Volume 1 — The First Four Days* is a mandatory read for you. *The Science of God* takes you through the thought process to enable you to speak intelligibly about Creation, the cosmos, evolution, and astrophysics.

Search: The Science Of God Book
SayItBooks.com

Announcements

Discover the Building Blocks of Life

Are you feeling confused about how plants really came to be? Did they evolve, or did some "God" Create them? Could they exist without the sunshine? When did gravity begin, and does that matter regarding the arrival of plants? If a God did Create the plants, then exactly how might that have occurred? Or if the plants evolved, then what did they evolve from?

There are many questions that need to be answered, but who has the time to study these things in depth? After all, scientists do this as a fulltime career and even they lack many answers to such questions.

The Science of God Volume 2 - Gravity, Land, Seas, and the Evolution of Plants offers unique perspectives to assist in quickly discerning the onslaught of information from both the religious and scientific sides of this debate. While there are some scientists and religious people who attempt to stand on both sides of the evolution versus Creation discussion, doing so often harms their credibility due to conflicts in their logic.

The Science of God Volume 2 - Gravity, Land, Seas, and the Evolution of Plants stands alone in explaining and answering the central questions that many people have surrounding the topic of plant evolution versus Creation.

Search: The Science Of God Book Volume 2
SayItBooks.com

Announcements

A Fishy Flying Account Crawling Out of Nowhere

Have you been trying to share your views in the evolution versus Creation debate but are thwarted at your every utterance? Are you reluctant to speak up and share your opinions because you're not sure what is or is not true? Sometimes we might even wonder if we should even bother pondering these things at all since no one can ever truly prove their theory to a point of it being "undeniable fact".

Take heart because there are more possibilities than are offered by most people on either side of the discussion. Bystanders often observe the views from both sides of the debate and will then consider those perspectives and try to balance them using logic, but we often fail to achieve that logical balance.

Balance is achieved by many people, but it is typically compromised in order to arrive at an agreeable viewpoint. Ignoring facts in this way is no way to discover truth.

The Science Of God Volume 3 – The creatures – Revolution or Evolution will not force you to ignore any true facts, and will guide you on your quest to see the clear path to how creatures came to be. God or Evolution? You decide, because everyone is welcome in the discussion!

**Search: The Science Of God Book Volume 3
SayItBooks.com**

Announcements

The Cornerstone of Moral Civilization

Was Jesus really the "Savior"? Did Noah really save humanity from extinction? Did Adam and Eve really get evicted from the Garden of Eden? And what does the word "Bible" mean anyway? When studying or even just reading the Bible, many questions arise to a point where the Bible can be confusing. But when you have certain information before you begin reading, it can instantly propel you to a deeper level of understanding by nothing more than knowing a few key points.

It takes people years to realize some of this information, yet it's not some big secret that only scholars and theologians know. No, this information is for everyone and it's easy to grasp these pieces of information about the Bible and some of the events described within it. Be prepared to have your current views challenged because many things are not as we have been taught.

To truly Understand the Bible, we must open our minds and toss aside all of our biases. Knowing and grasping the often-unrealized basic information presented in *Understanding The Bible - The Bible How-To Manual and The Things We Don't See* brings the Bible to life in a way that shows you, personally, its undeniable relevance to the world, to our culture, and to your very own life!

Search: Understanding The Bible Book
SayItBooks.com

Announcements

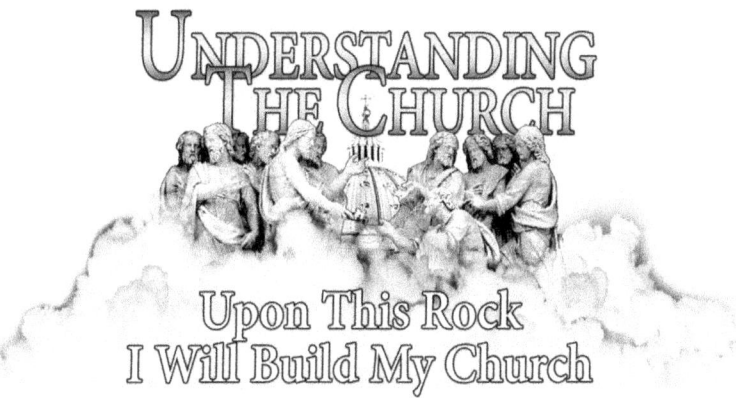

Church in the Lurch - a House Built Upon Sand

The Church is rapidly dying, and much of the clergy in recent times have been doing it more harm than good. People are fleeing from the Churches as they seek a religious perspective that fits a modern worldview. Should we revive this old Church and try to save it from its own demise? What exactly is "The Church", and who or which of the many religions is the official caretaker of it?

The Christian religions of the world have done their fair share of damage to themselves and to the world, but in the bigger picture, they have done more good than damage. Saving the Church is probably worth our collective efforts because the Churches are perhaps the most charitable group of organizations that existed throughout history and even up to today.

The main reason that the Churches are in the rough condition that they are today is due to a lack of understanding by clergy and congregation. We can overcome this dark era of the Church and revive it only through *Understanding The Church*.

Understanding The Church will help you in Bible study, or even to simply better understand the Church. But most importantly, *Understanding The Church – Upon This Rock I Will Build My Church* will help to revive this dying patient.

Search: Understanding The Church Book
SayItBooks.com

Announcements

Rock the Boat with Layers of Truth

Do you believe that the entire world flooded roughly four thousand years ago and that a man named Noah built a large boat to save a small remnant of human and animal life that would repopulate the entire Earth? This is the belief of many Christians, Jews, and Muslims, but then we have those who believe that the entire story was written thousands of years ago for entertainment only.

Could either case be true? Is either realistic? After all there is a lot of evidence of catastrophic worldwide flooding. But then there are those making the point that there's not enough water on Earth to cover the mountains. So, which, if either, is it? If either case were proven to be undeniably true it would have major impact on opposing perspectives. If it never occurred, it would devastate most Bible-based religions. But how would it affect modern sciences if it was proven true? It would force every scientist to face a reality for which they have not been educated.

Take a journey through these and other Biblical flood questions and consider the perspectives presented in *The Science Of God Volume 5 – Boats, Floods, and Noah – The Deluge*, a truly logical scientific explanation of the viability regarding the Biblical flood of Noah's time.

Search: The Science Of God Book Volume 5
SayItBooks.com

Notes

Notes

Notes

Notes

www.ingramcontent.com/pod-product-compliance
Lightning Source LLC
Chambersburg PA
CBHW071702090426
42738CB00009B/1636